THE

CHEETAHS

OF DE WILDT

THE
CHEETAHS

OF DE WILDT

ANN VAN DYK

STRUIK

TO
GODFREY
WHO MADE A DREAM
COME TRUE

Struik Publishers (Pty) Ltd
(a member of The Struik Publishing Group (Pty) Ltd)
Cornelis Struik House
80 McKenzie Street
Cape Town 8001

Reg. No.: 54/00965/07

First published 1991
Second impression 1995

ISBN 1 86825 187 X

Edited by Phillida Brooke Simons

Reproduction by Sparhams Cape, Cape Town
Printed and bound by Kyodo Printing Co. (Singapore) (Pte) Ltd, Singapore

CONTENTS

ACKNOWLEDGEMENTS

So many friends, conservation bodies, and businesses in the economic sector have helped me at De Wildt during the past years that it is impossible to list every one of them. I am most grateful to them all for their interest and commitment.

My deep appreciation goes to Professor Dave Meltzer, Faculty of Veterinary Science, University of Pretoria, Onderstepoort, for the major role he has played in the establishment of De Wildt. He is a friend to whom I shall always be indebted.

To the following I extend special thanks: past and present councils, directors and staff of the National Zoological Gardens, Pretoria; Professors Brough Coubrough, Henk Bertschinger, their colleagues and staff, Faculty of Veterinary Science, University of Pretoria, Onderstepoort; Professor John Skinner, his colleagues and staff of the Mammal Research Institute, University of Pretoria; Professor Joseph van Heerden, his colleagues and staff of the Medical University of South Africa, Medunsa; Dr Piet Mulder and staff of the Department of Nature Conservation, Pretoria; Dr Anton Rupert and the South African Nature Foundation, Stellenbosch; Dr John Ledger of the Endangered Wildlife Trust, Johannesburg, and the Vulture Study Group, Johannesburg; Mr Tony Ferrar and the Wildlife Society of Southern Africa.

My sincere gratitude is extended to Dick Reucassel for his friendship, considerable assistance and excellent photography; to Cynthia Kemp for her invaluable advice concerning the manuscript – without her encouragement this book would never have been written; to Alan Strachan for aid with photographs and for the maps and diagrams reproduced here; to Caroline van Zyl and Eugene Marais for typing the script; to Miriam Meltzer for her help and expertly conducted tours at De Wildt; to Denny and Vi Crowther for their interest and enthusiasm; to Hennie and Koba Diederiks whose friendship has known no bounds; to my team at De Wildt whose selfless and untiring efforts through the years have helped to make the De Wildt Cheetah Research Centre what it is today; and of course to my family, especially Philip and Michael, whose continued love and support means so much to me.

I acknowledge all those who have so kindly given me slides and photographs. Some, whose pictures appear in this book, I have not been able to trace and I accordingly make no claim to copyright in these cases and hope it will be possible to name the owners in any future reprints.

FOREWORD

For almost two decades Ann van Dyk has waged a relentless campaign for the conservation of the cheetah. This book is largely about that campaign and about the dedication and purposefulness which eventually was to bring success. Ann's account could have been a real tear-jerker because she became deeply and emotionally involved in the cause she was expounding. However, it soon becomes apparent that emotions have their place in the story of the De Wildt cheetahs. Moreover, Ann's account of achievement is much too modest.

While this is yet another record of achievement in the field of nature conservation, it differs in that Ann describes what so many of us have dreamt of doing. It is the story of a major contribution being made to conserving what was once an endangered species, the cheetah. Despite the immense effort that was made by Ann and her brother Godfrey in terms of both time and personal expense, they became so dedicated to finding a solution that the heartbreaks which always accompany projects such as these never deterred them. It is inevitable that there are heartbreaks which remain indelibly imprinted on one's memory and many of these are described in this book. Yet the successes were far greater, for they led to the unlocking of the key to captive breeding of the cheetah. This was a wonderful breakthrough, for it has meant the removal of the cheetah from the endangered list and there is now a supply of cheetahs that can be sent to zoological gardens throughout the world. Also, for the first time there is now a recipe for breeding these cats in captivity.

The second important impact that Ann's campaign has made, is that it is now an accepted fact that it is not easy to reintroduce captive-bred cheetahs into the wild in areas where non-captive cheetahs already exist. She describes in detail attempts to do this and why it is not feasible. Only in specially selected areas can such a project be successfully undertaken.

The third secret unlocked has been the identification and breeding of the so-called 'king' cheetah, proving once and for all that it is not a separate species: the coat of the animal which has a dramatically different colour pattern from that of other cheetahs, is but an aberration so common among felids. But what a handsome animal it is, and surely now it will be exhibited more widely than ever before.

While the cheetah project was the base from which Ann launched her conservation ethic, it soon widened to include other endangered animal species, such as brown hyaenas, wild dogs, servals, suni antelope and riverine rabbits. All of these have been successfully bred for later redistribution in the wild, thus helping to repopulate areas where such species have disappeared or are no longer abundant. In time this conservation biology role will become more and more important as it becomes necessary to build up and maintain adequate gene pools.

Ann has a long history of collaborating with scientists in all her projects, and it is this teamwork that has contributed to the success of projects launched under the auspices of the National Zoological Gardens. This does not detract one iota from the enormous contribution she herself has made. She has set a wonderful example which is now recorded for posterity.

PROFESSOR JOHN D. SKINNER
Head of the Department of Zoology
and Director of the Mammal Research Institute, University of Pretoria

Introduction

A soft rosy haze lay on the eastern horizon. Another hot day was dawning and despite the long cool night, the dust of the previous day had still not settled on this quiet north-eastern corner of the Transvaal. Typical bushveld – tall dry grass and areas of dense bush – stretched as far as the eye could see. The silence lay like a blanket over the scene, broken only by the wail of a jackal or the distant whoop of a hyaena. Suddenly, as if at a given signal, a bulbul broke the stillness with a noisy birdcall and started off the other birds in their chorus heralding the new day.

In the farmyard the cattle stamped noisily in the kraal, the young calves restlessly nudging their mothers, eager for their first warm drink of the day. Carrying a shotgun, the farmer emerged from his homestead and walked purposefully towards his livestock. For the past few weeks an unknown predator had taken a number of his valuable calves and he was determined to put an end to this unnecessary financial loss. After releasing the cows with young to graze on the open veld, the farmer took up his position behind an old ox-wagon and waited. Suddenly, in spite of the hazy air, his well-trained eye caught the movement of a long sleek form stealthily creeping up through the dry grass. Completely motionless, muscles taut, the shape stopped for a moment. The predator's eye settled on a calf that had wandered a fair distance from its mother. Perhaps it was the pangs of her own hunger and those of her three young cubs hidden not far off in the tall grass, that made the cheetah – for that was the predator – unaware of her surroundings for she did not see the farmer raise his gun above the wheel of the ox-wagon. A single shot rang out over the valley. The cattle stampeded and the dark form collapsed in the grass and was still.

'This time I have you, you bastard!' the farmer cried out ecstatically and ran to the spot where the animal lay motionless.

His joy, however, turned to consternation when he realized that the cheetah he had just killed was a lactating female, her teats swollen and heavy with milk. For the moment forgetting his stock losses and anger, he thought of the young cubs hungrily awaiting the return of their mother. They would never have known their father as he, like all cheetahs, would have left the female immediately after mating, making her the sole protector of her young.

Calling to one of his farmhands, the farmer instructed him to bring out the dogs on chains, and with them he set off to scent out the cubs. The young cheetahs could not be far away, he surmised; they should not be left to die of hunger.

Within an hour the dogs had sniffed out three spitting and snarling youngsters. Lifting them one by one by the scruffs of their necks, he placed them together in a sack. 'I must take them to my wife,' he shrugged at his farmhands. 'She'll know how to feed them.'

One of the three cubs turned out to be a weakling; it refused to eat and died the following day. But the remaining two proved to be strong and healthy. Both were females, and one was duly named 'Kromstert' as she had a kink at the tip of her tail, while the other was 'Lady' because she was seldom aggressive and remained impassively aloof.

A few days later the farmer chanced to read a report in the local newspaper of a project that was to be launched to breed cheetahs in captive conditions. Prompted by the dwindling number of these large cats, a decision had been made to establish a research and breeding centre at De Wildt, a farming community on the northern side of the Magaliesberg range, some 40 kilometres west of Pretoria. Under the auspices of the National Zoological Gardens in Pretoria, an appeal was made to farmers to attempt to catch rather than shoot cheetahs that were proving a problem for them. The zoo undertook to arrange permits for these animals and their transportation to the centre.

Realizing that the two fluffy and mischievous cubs his wife was now rearing would one day become both difficult to feed and a danger to his livestock, the farmer quickly made a long-distance telephone call and within a week Kromstert

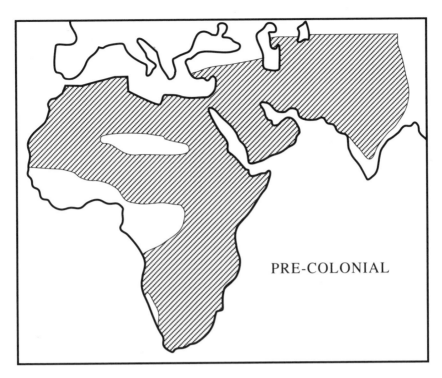

PRE-COLONIAL

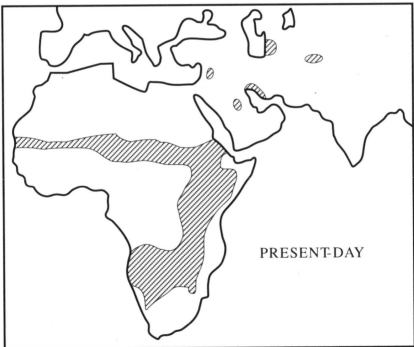

PRESENT-DAY

Figure 1   Estimated distribution of the cheetah

and Lady were on their way to a new home. Unknown to them, they were about to take part in a unique experiment which set out to help save the cheetah from extinction.

From the early 1960s the cheetah, *Acinonyx jubatus*, had been regarded as an endangered species. Whereas in the distant past this swift-footed long-legged cat roamed through nearly all of Africa, Arabia, and what is today Iran, Afghanistan, Pakistan and India, at present its range is reduced to specific areas in Africa and a few small pockets of land in the Asian countries. Apart from the fur trade, the cause of this decline in numbers is mainly due to the fast-expanding human population that, together with agricultural development, has rapidly taken over age-old cheetah habitats. One of the earth's fastest mammals, this big cat, unlike the leopard, cannot exist in mountainous and thickly forested areas. It needs the expanse of flatland or open plain on which to hunt and run down its prey. This flatland is coveted both by man – for crops and livestock – and the cheetah.

Added to this, unlike most other felids or cats, the cheetah is difficult to breed in captivity. With this in mind the De Wildt Cheetah Research Centre was launched in 1971 by Dr Frank Brand, then director of the National Zoological Gardens in Pretoria. For him it had been a great challenge: could anyone ever unravel the secrets that pertained to the procreation of this stolid and regal cat or was it destined to join the dodo and the dinosaur, to become a has-been, a skin upon a wall or a stuffed specimen in a museum? This then is the story of the De Wildt Cheetah Research Centre, and of its team of helpers who took up the challenge – that of preserving the existence of a cat species.

CHAPTER 1

The Early Days

*I*n the late 1940s my parents, Godfrey and Catherine van Dyk, both medical doctors in Pretoria and nearing retirement age, decided to buy a retreat in the country. This, they reasoned, would enable them to escape from the pressure of city life during weekends and to own a place where they could grow vegetables, raise chickens and enjoy the pleasures of country living. A medical practice is demanding; they had both lived strenuous lives and they had raised a family of five children. Margaret, the eldest, was happily married; then followed Godfrey and James, both at university studying medicine. I was the next to see the light of day and lastly came Reginald, my youngest brother. Reg and I were then in our teens and still at school. The thought of my parents buying a farm had always been a dream of mine: the discipline of school and bookwork had never come easily to me, and while living in Pretoria I would slip away from my school homework whenever possible and attend to the flowers and vegetables I had planted in our large garden. Now I imagined cows to milk, chickens to attend to and fields of cut flowers and vegetables to be marketed. At that age, how easily my imagination ran away with me. Farming seemed so straightforward and so profitable. After the seeds were planted, watered and cultivated, you simply reaped the crop and – considering the prices my mother had to pay for vegetables – from the proceeds we would make a good living. Selecting a farm was a marvellous excuse for me to ignore homework, and with my mother I would search through the 'Farms for Sale' columns of the local newspapers. Early on I learnt the meaning of 'fantastic offer'. Invariably it meant a dejected drive back home, all of us wondering what the definition of 'ordinary' must be.

It was with cynical eyes that we read the advertisement: 'De Wildt Estate – the chance of a lifetime! Due to ill health the owner of this productive 80-morgen farm is forced to sell. Fifty morgen of the land is koppie, covered with natural bush and grass, while the remaining 30 morgen are under intensive cultivation. With a sub-tropical climate that favours the existing avocado pear, mango, early peach and apricot trees, all now in full bearing, it is warm enough to grow winter flowers and vegetables. Nestling in the northern foothills of the Magaliesberg mountain range, 42 miles north of Johannesburg and 25 miles west of Pretoria, it is ideally situated for the man who wants to live in the city and own a lucrative farming operation.' The advertisement ended with the statement that there was a competent farm manager to boot. Despite uneasy thoughts that this could be just another pipedream, this farm did seem to be different and promised to fulfil our expectations. Here we would all be able to spend relaxing weekends and then simply collect the cheques from the produce and bank them at the end of each week. We didn't have to do much soul searching, and there was little need for family discussion. Instinct told us that it was the right place for us, and it was not long before the paperwork and transactions were out of the way. On 13 September 1950 De Wildt Estate became ours.

The Magaliesberg was then one of the last remaining areas close to the sprawling cities of Johannesburg and Pretoria where nature had not been tampered with. The mountain range stretched from east to west for approximately 200 kilometres and rose approximately 450 metres above the surrounding countryside. The rugged and sun-baked northern slopes were covered with rocks which were interspersed with tufts of corn-coloured grass and clusters of indigenous bush, the surface being broken by deep gorges, dropping steeply down to underground streams which bubbled up to the surface after heavy rains. From the moss and fern-covered walls of the ravines, tall moepel, wild fig and white stinkwood trees vied with one another for a place in the sun. Below these slopes was a flat plain with rich soil fed by mountain springs. It was here, on level ground, that farming activities took place, leaving the rocky area above to nature.

Bounding our newly acquired farm to the north was a road that led eastwards to Pretoria and westwards in the direction of Brits, the latter being the centre of a rich farmland area watered by the Crocodile River and its Hartbeespoort Dam irrigation scheme. Nineteenth-century hunters had taken care of the huge herds of animals that once inhabited the Magaliesberg and its surrounds, but many species had nevertheless survived – antelope such as the agile klipspringer, the small well-camouflaged grey duiker and the large mountain reedbuck that grazed on the mountain slopes. Dassies – or hyraxes – abounded in large numbers and were often seen early in the morning emerging from rock crevices to warm themselves in the first rays of the sun. Troops of baboons and vervet monkeys, seldom seen but often heard, combed the range in search of food. They found sustenance in the form of young and tender shoots, as well as in seeds, berries and wild fruit, and supplemented this diet with tasty morsels such as eggs, nestlings, scorpions and insects. I could hardly believe that we now owned a part of this unspoilt terrain. I was ecstatic. On leaving school I had been unable to decide on a career and hated the idea of an office job, but now I could join my mother in a farming enterprise.

Soon, however, my parents found that a disturbing element had crept into their lives. Almost every day, at the end of the list of patients to be seen at the consulting rooms, was the same name – that of the manager of the farm. At each visit he requested authorization for payment of some kind: fuel or fertilizer or spares or wages. It all seemed so out of character with the idyllic advertisement. For the first five years our farming venture seemed a bottomless pit of never-ending bills. We were plunged into even deeper despair when one day a market-agent visitor solemnly pronounced that De Wildt Estate had never paid and never would. Only the challenge and eagerness to succeed kept us going. Later, when we discovered that our 'competent' manager was dividing the crop in half – one half for us and the other for himself – he was politely asked to seek employment elsewhere.

After this my parents decided to move their home from Pretoria to the farm, stubbornly refusing to give in so soon. Despite his earlier plan of retirement, my father knew that he would have to continue with his medical practice in order to help financially until the farming operation could pay for itself. My mother held the purse strings and, being a true Scot, thriftiness was the password. I shall never forget her oft-repeated expression, 'Keep your nose to the grindstone!'

Not long after we had moved to the farm, Godfrey my eldest brother, then in his third year of medical studies at the University of Cape Town, informed my parents that he was not happy in his medical career and that he wished to join us on the farm. This did not altogether surprise my parents as Godfrey had always had an interest in farming and whenever possible had spent his vacations on a friend's cattle ranch in the Western Transvaal. He was very practical and mechanically minded and better suited to a farming career.

Godfrey was welcomed with open arms, and the seeds of a prosperous farming operation were sown. Young and energetic, and now happy in his new-found work, he became the driving force behind a team of faithful workers. When thinking of them, memories of people like Joel Tollo, Moses Phiri, Piet Banda, Aston Phiri and David Maleti flash through my mind. I remember one or other of them proudly presenting my mother with the first peach pickings or a bunch of early winter flowers. Life proved to be hard for everyone during those first years on the farm, but they were good times too. Three thousand peach trees had to be pruned; there were onions to be reaped, tomatoes to be boxed, and flower bunches to be tied. The days were never long enough and often we were obliged to work well into the night.

While Godfrey saw to the planting and reaping, I worked with a team of women in the packing shed. They affectionately called me 'Mamabroek' because I always wore jeans. Even today, many years later, a cheery face from those past days will greet me in the De Wildt area and I will think back to those times with nostalgia. And so, slowly, the farm developed into a viable concern.

The years passed and after both my parents had died – my father in the late 1950s and my mother in the early 1960s – Godfrey and I were in sole charge of the farm. It was then that we decided to replace the crop of perishable peaches and apricots with oranges and to venture into an intensive egg-farming operation. We started the latter in 1964 with 500 hens and within 10 years had built up a prosperous business which housed 100 000 cackling red-faced ladies who constantly demanded food and attention.

In those early days many a stray animal, wild and tame, had found shelter in our home, mainly due to my concern for them. There was Mdala the horse, Piet the mule, Varkie the pig, Stompie the hyrax, Kimba the puma, and Dingo the Australian wild dog – besides household cats and dogs too numerous to name. They all wanted affection and friendship and, despite their numbers and our farming commitments, they all received our love. Mdala (meaning 'old') and Piet had together wandered onto the farm one day. Lost and dejected, they had looked a sorry sight: merely skin and bone, their bodies had been covered with sores and parasites and their coats were dry and matted. As both animals were extremely hungry and thirsty, on Godfrey's advice I had given them water and fed them wheaten bran, a little and often. They had shown no aggression as I stroked them, but the dull sunken eyes told a very sad story. After a few weeks, on good grazing and a high protein supplement, their condition had improved beyond recognition. Their coats began to shine and the light returned to their eyes. In the district word went out that we had found a stray horse and mule, but the pair was never claimed and Mdala and Piet made no attempt to leave. Soon the farm became known in the neighbourhood as the local dumping ground for stray, sick and wounded animals.

It was amazing to see what great patience Godfrey had with these hapless beasts and the hardships he went through because of my love for animals. I smile now as I think of one occasion when, while dressing for a business trip to Pretoria, he found that Dingo had chewed the backs off his best shoes. The air had become quite electric with Godfrey swearing under his breath and so I had discreetly disappeared. But on his return from the city that evening, all was forgotten when an excited Dingo rushed to greet him. I also noticed through the corner of my eye that Godfrey was wearing a pair of brand-new shoes!

Stompie, the hyrax, was a great character and an unusual pet. When very small he had been brought to us by one of our workers as he had a festering open wound on one of his hindlegs. As hyraxes are vegetarians we had fed him on leaves from indigenous plants, but in addition to this diet he loved brown bread and biscuits and was very partial to the occasional hard-boiled egg. When fully grown, Stompie had the appearance of a large rotund guinea-pig. In no time he had become a great favourite with Godfrey. In the early evening, after a gruelling and hot day on the farm, Godfrey would relax with a sundowner on the veranda of our home. Stompie would then suddenly appear, hop onto Godfrey's shoulder and drape himself round his neck, announcing his arrival with twittering noises. Thereupon Godfrey would pour a scotch-on-the-rocks into Stompie's own special glass and the two would quietly sip their drinks together, enjoying the peace and serenity. Stompie was a great lover of music and a composer too. After dinner he would jump from Godfrey's shoulder onto the piano until the keyboard cover was opened for him. Then he would run up and down the keys apparently fascinated by the noise that his feet were able to produce. Like all hyraxes, Stompie was a great sun-lover and revelled in the heat. During the winter months, while we relaxed at night in front of a crackling log-fire, Stompie would be there too, stretched out on his stomach on the rug, twittering contentedly. Later when we retired he was put to bed on a hot-water bottle in his own wooden box.

The only animal in our household that Godfrey and I somehow never got close to was the puma, Kimba. Also known as a 'mountain lion' or 'cougar', the puma is foreign to southern Africa but still roams wild in many parts of North and South America. Similar to but much smaller than the African lion, it is a powerful cat to contend with, being strong, cunning and agile. Kimba was taken in as a boarder for a few years and during that time he merely tolerated us for he had only one master – his owner who lived in Pretoria and who could not keep the animal in the city. During the long winter evenings as we sat in front of the open fire, half a puma face would appear, peeping at us from behind the door. Having scanned the room for intruders, he would then wander in, dispassionate and aloof, and flop down at our feet ignoring us completely. Perhaps he held us responsible for the absence of his master. Whatever it was, we could never break down the invisible barrier that separated us.

One day Kimba disappeared. A visitor had inadvertently left the garden gate open and Kimba had taken the opportunity to regain his freedom. Together with a team of farm workers we searched the rugged Magaliesberg range for a solid week. It seemed obvious that Kimba would seek the safety and cover of the mountain as this was an ideal place for an animal to hide, and here food was to be found in abundance. If need be, Kimba could live on birds such as the plentiful guinea-fowl and francolin. During the search for our boarder we left our telephone number with all the local farmers and offered a reward for Kimba's recovery. We were cheered by reports from a few of the locals that they had actually seen him, though we were disappointed that they had then fled from him, thinking – so they said – that he was a lion. Then one Saturday afternoon as I was bent over the washbasin with my blonde hair a mass of white soapsuds, the telephone rang. A neighbour's voice rasped: 'Ann, come quickly – your big cat is in my duck enclosure!'

With soapy bubbles dripping down my neck and face I swiftly turbanned my head in a towel and dashed to the next-door farm. There was His Lordship, as arrogant as ever and without a care in the world, feasting on a fat muscovy duck. At the sight of him, all consideration for his regal feelings was abandoned and I roughly picked up a most indignant puma and threw him onto the back seat of my car. It was with sighs of relief that we responded to the information received shortly afterwards that Kimba's owner had bought a small game-farm and that he could be returned to his master.

Dingo was quite the opposite of Kimba. I had been given him as a pup and he had grown up together with Astra, our German Shepherd. They were inseparable. Despite his gentleness and affection for us, on reaching maturity Dingo's wild instincts seemed to get the better of him. He would disappear from our house at night and while out on the prowl would kill any cat in sight and often raid the neighbours' chicken houses. Fearing that he would be shot, we were obliged to confine him to the small fenced enclosure that surrounded our house. Strangely, one cat, Cookie, never fled when approached by Dingo and the two became great friends and slept together on my bed at night. This seemed to indicate that the predator's instinct to chase is heightened by the hunted turning its back in flight.

One Sunday, a peaceful lazy summer's morning, Godfrey beckoned to me and, waving a weekly farming magazine, said, 'How would you like to own a pair of cheetahs?'

Two arrogant yet appealing faces on the cover of the magazine seemed to look right through me. 'I'd be scared,' I replied, though I was inwardly envious of the owner who stood as a blur in the background of the photograph.

'I think they'd make most interesting companions,' mused Godfrey. 'By the way, talking of wild cats, I think you should do something about the population explosion on this farm!'

He was right. What had started off as an innocent daily feed of mealie-meal and cracked eggs for a few stray feral cats, had turned into free board and lodging for well over 100. Our thoughts were turned from the cheetah photograph to our own cats and we hastily decided to spay every female. Our farm workers were rewarded for every one caught and the local vet had visions of early retirement on the proceeds of spaying fees. It was our first taste of the expense of farming cats.

Some two years later, in October 1968, a local farmer who had learnt of our interest in wild animals, passed on to us the information that two cheetah cubs were being offered for sale by a farmer in the Northern Transvaal. Discussion between me and Godfrey was unnecessary – it was simply a matter of arranging to fetch the cubs as soon as possible. We had heard that permits were needed to enable us to keep indigenous wild animals, but decided to apply for them once the cheetahs were in our care.

I remember well the morning that Godfrey, together with farmhand Joel Tollo, set off early to fetch our cheetahs. The back seat of the car had been removed to make room for the crate in which the cubs would be transported. Having started work before dawn, I found that day to be one of the longest I had ever experienced. The egg packing kept me busy, but my mind was not on it; I kept imagining what Godfrey was doing, where he was, and what the cubs would be like. At last the nose of the car turned slowly into the long sandy driveway that led up to the farmhouse. I barely greeted Godfrey in my excitement and hardly waited for the car to stop before peering in through the back window. Two aggressive yet appealing cheetah faces looked up at me. Despite a flicker of apprehension, I was beside myself with joy and quickly set about making my two new charges feel at home.

A room for the cubs had been cleared in the house as we felt that they would be safer living near to us while at the same time we would get to know one another far more quickly. Although they were still spitting and snarling angrily, we managed to manoeuvre them into their room where, after they had devoured the big bowl of minced meat that was offered to them, they settled down for the night, very close to each other and on an old rug. This was something I was to learn later: young cheetahs brought up together, though not necessarily litter mates, have an exceptionally strong bond and even when separated for a few years will, upon reunion, recognize and greet one another as if they had never been parted.

The morning after the cubs' arrival I opened the door of their room and then proceeded to breakfast with Godfrey. It was not long before there were sounds of a scuffle in the passage and a minute later two faces, one above the other, peered round the doorway. We purposely paid no attention to them and soon the cubs were sniffing about the dining-room. At any sudden or unusual sound they would scurry back to their room. It took only a couple of days before they knew their way about the house and garden, though they were still wary of Godfrey and me. It seemed that their greatest delight was to sit together on a large armchair, their frontlegs dangling over the headrest, and to watch through the adjacent window the activities of the farm outside.

On enquiring about the necessary permit to enable us to keep our charges, we found to our dismay that we had broken the law. In the province of the Transvaal, we were told, it is illegal to purchase and transport cheetahs without permission. No indigenous wild animal is allowed to be kept in captivity without a valid reason and the authorities had no option other than to confiscate our cubs. We had been prepared to look after them as we did our domestic pets and to give them all the attention they needed, but it was only years later that we understood the reasoning behind this nature conservation legislation and in time we came to support it fully.

One may have heard of young wild animals, such as baboons, monkeys, jackals, antelope and others, being caught or taken away from their mothers. 'Fantastic pets for the kids,' is a common remark that one hears, but unfortunately these 'cuddly little things' grow, and on reaching maturity almost invariably become aggressive, unmanageable and dangerous. As adults, they cannot then be released back into their natural habitat as their own kind will not accept them, so more often than not the only answer is euthanasia. We ourselves, sadly, had been guilty of transgressing the law but fortunately in our case our charges were to be spared and would be cared for by the National Zoological Gardens in Pretoria. We watched dismally as the Nature Conservation truck drove off to the zoo the next day, taking with them 'our' cubs.

Knowing full well that there was no way of reversing the confiscation, but aware of my sadness and disappointment, Godfrey decided to pursue the possibility of legally keeping cheetahs. Our sister, Margaret, lived in Pretoria and she arranged for us to meet a friend of hers, a well-known wildlife enthusiast and member of the board of trustees of the zoo in Pretoria. With charm and understanding, our minds were taken off the subject of the impounded cheetahs and we were told instead of a project that the zoo had had in mind for some time. Aware that cheetahs were on the list of endangered animal species, that they rarely bred in captivity and that their numbers were declining rapidly, the zoo authorities had thought of acquiring a large tract of land in which they could introduce cheetahs and, through a study of their behaviour and habits, could attempt to breed them in semi-captivity.

Our citrus trees and poultry sheds were situated on the flat arable part of the estate and immediately behind them rose the unused rocky, wild and unproductive part of the farm, the foothills of the Magaliesberg range. There the indigenous vegetation had never been tampered with by man and had been left to the care of nature. With 40 hectares of hillside available, we decided to offer the use of the ground to the director of the National Zoological Gardens in Pretoria, Dr Frank Brand, if he thought it would be suitable for the proposed cheetah-breeding project – or any wild animal breeding project for that matter.

Because of urgent orders for eggs, I could not accompany Godfrey on the morning that he set out for Pretoria to discuss our suggestion with Frank Brand. I disciplined myself not to be too excited, remembering the setback of our earlier encounter with cheetahs, but on this occasion it was Godfrey who was unable to contain his enthusiasm. The car had barely halted when he was already relaying to me the zoo's interest in our property and its willingness, if the area was found to be suitable, to attempt to breed cheetahs on our farm. Frank Brand and Dr Hannes Koen, the chairman of the zoo's board of trustees, were to visit us as soon as possible to inspect the land that we had suggested using for the project.

I could not believe what I was hearing. It seemed that my wildest dream was about to come true.

CHAPTER 2

A Dream Come True

*T*wo weeks later, in November 1968, we guided Frank Brand and Hannes Koen up the koppie on De Wildt Estate. Looking south from the top of this rocky hill, we could see the northern slopes of the nearby Magaliesberg range, and to the west an area of level ground with a small stream, fed by a natural spring, winding through it. Godfrey pointed out that the vegetation on this land consisted entirely of natural bush without any exotic or foreign plants. Tall grass grew between scattered clumps of acacia thorn bush, large marula, wild syringa and red ivory trees. Branches of giant wild fig and moepel trees hung over the meandering stream, and in the cool shadows of their boughs we all paused to watch the antics of a group of vervet monkeys which had come down from the mountain in search of food. We halted again when we came to our favourite tree, named by us the 'Big Tree'. It was an enormous moepel or Transvaal Red Milkwood, estimated to be at least 300 years old, and in its huge spreading branches we showed our visitors a large untidy platform of sticks, the nest of a hadeda ibis. I have often imagined how, before man's encroachment, a leopard might have silently watched and waited here, in the dense riverine bush, for an unsuspecting animal to drink at the stream below. After a short, swift and deadly lunge at its prey, the leopard may have hoisted the dead animal onto the large moepel tree's branches. Here, hidden from scavengers, it would have had sufficient meat to satisfy its hunger for a few days.

We explained to our guests that, although some animal wildlife still found refuge in the actual Magaliesberg range, very little remained below in the valley. Baboons were heard in the mountains during the day, jackals called at night and occasionally a leopard was encountered – often only to be hunted and shot by stock-owning farmers. Fear of predators always seemed to result in the killing of these leopards, even though most cats prefer to give way rather than to attack.

Although the area was seen to be perhaps a little too stony, everyone was enthusiastic about our offer. After lengthy discussions and a good deal of planning, a non-paying lease was drawn up stating that the zoo could, for 15 years, make use of the 40 hectares of land for the breeding of cheetahs. The animals would be provided by and remain the property of the zoo. The latter would be responsible for supplying the cheetahs with food and for any veterinary care. Godfrey and I were appointed honorary custodians of animals sent to the farm.

And so, slowly, the De Wildt Cheetah Research Centre's project was born. As we owned the property we were obliged to erect the necessary buildings and fences at our own expense. We realized only later that this was no mean financial responsibility for it involved three kilometres of diamond-mesh fencing which had to be three metres high with an overhang of one metre for added security. At the base, the fences had to be embedded in 40 centimetres of concrete to prevent wild jackals or farm dogs from burrowing beneath them.

In early October 1969 the first fencing poles were planted. Work on the farm now developed into a joint venture for Godfrey and me. After completing the dispatching of egg orders, usually by midday, Godfrey and a team of workers would dig the holes into which the poles would be fixed in concrete. A road had to be forged up the koppie and the farm's womenfolk and I were assigned to this task. These were cheerful times despite the backbreaking work: the men would laugh and joke, chasing one another on to keep up with the concrete mixer; the women sang in unison as we cleared the roadway of bush and stone. Many of our workers had lived in the northern, less industrialized areas of the country, and knew the cheetah. They could not understand why we were trying to save, let alone create more of an animal that they knew as a killer of precious livestock.

The months went by and we learnt that fencing on a rocky koppie was not without problems. One area of the hill was made up of large granite boulders. Like an iceberg, very little of the rock showed on the surface and in the end a tractor often had to be used to coax out the offending unseen two-thirds. At times, when there were shortages of concrete, steel, and diamond-mesh fencing, we were obliged to lay down tools and endure the frustration of waiting, not knowing when supplies would again be available. After more than 12 months of gruelling labour, the erection of the fences was still not complete and it seemed a never-ending task. Years later the fencing would all be taken for granted.

The feeding of the cheetahs had to be catered for, and to this end an empty room adjoining the egg-packing station was converted into a butchery. There we installed a large deepfreeze cabinet in which meat would be stored, as well as several stainless steel tables for the preparation of the food. The zoo would obtain the carcasses – cows, horses, donkeys, whatever was available from abattoirs – and undertook to send supplies to us. Any suitable dead hens from the egg-farm would also be used to feed the cheetahs.

We were riding on the crest of the wave and we knew it. The fencing for the cheetah enclosures, or camps, was progressing well; on the egg-farm we had just been able to buy a modern imported egg-grading machine; and at home plans had been drawn up to alter our somewhat dilapidated farmhouse. So many long-wished-for dreams were about to become reality.

Then overnight, disaster struck in the form of Newcastle disease. Until the early 1970s it had been virtually unknown in southern Africa; after it was introduced by the imported breeding-eggs of a new hybrid strain, it became the dreaded scourge of poultry farmers. On hearing of the outbreak of the disease, we had taken all the necessary and recommended precautions. The farm gates were kept locked, allowing only the delivery trucks to enter and leave, and the wheels of these vehicles were always sprayed with a strong disinfectant before entry into the area where our poultry was housed. Every morning, after showering, our staff were issued with clean overalls and gumboots and they were forbidden to leave the farm until the end of the working day. We all accepted these austere measures out of necessity and were confident that in doing so we could isolate ourselves from the disease. However, apart from vehicles and humans, there was still the possibility that the virus could be carried from one farm to another by wild birds.

Suddenly, early in November 1970, a dramatic rise in the mortality of our hens caused Godfrey to take a few of the dead birds to the Veterinary Research Institute at nearby Onderstepoort for post-mortem examination. We knew that in the circumstances we had to be extremely careful of any changes that were noticed in our birds.

That day, just as I was sitting down to lunch, I was annoyed to hear the distant hooting of a car coming from the direction of the locked gates. I had been expecting a salesman and wondered why he had not bothered to walk up to the house as it was quite separate from the poultry area. I cursed inwardly and set off in the truck down the dusty road to the farm entrance. On seeing the official registration plate of the vehicle parked at the gates, my stomach turned. I greeted the driver who at once asked me whether he was on De Wildt Estate.

'Newcastle?' I blurted out.

'I'm afraid so,' was the man's simple reply. 'The dead birds we received this morning for a post-mortem examination were definitely infected with the disease.'

He hesitated, then followed with: 'I'm afraid I have to place you in quarantine. No eggs, poultry, or chicken manure can leave your farm until further notice.'

'But I have a truck packed with egg orders ready to leave at two o'clock this afternoon. Surely I can send them out?' I protested.

'I'm afraid not. Newcastle disease is something new and a good vaccine has yet to be found. We can't afford to take chances – the disease has already broken out on a number of farms.'

Knowing that others were in the same predicament was no comfort, and dejectedly I returned to the house. Not long afterwards, the farm dogs began to bark, announcing Godfrey's return from Pretoria. I knew him of old – he would be able to cope with this seemingly insoluble situation quite calmly. He would be unharassed and positive. I bombarded him with a stream of questions: 'What will happen now? How can we telephone the supermarkets, dairies and hospitals and simply say that for three weeks – perhaps longer – we won't be able to supply them with eggs? We'll have no income! What do we do with all the quarantined eggs and poultry?'

For a while Godfrey remained silent and then slowly he spelt out his plan. 'We have no alternative but to buy in eggs and so fulfil our orders. There is a market surplus available at the moment, but let's hope that the bank manager agrees to this proposal.'

Fortunately we were financially sound for Godfrey had always been cautious with money – the Scotch blood of my mother, I always thought. An amount had been carefully put aside for just such a rainy day and this was to see us through this troubled time and save us from ruination.

Soon it was all arranged. A friend living on a farm some 10 kilometres away allowed us the use of one of his large storerooms as a packing depot and it was from there that we carried on our business. Our customers seemed unperturbed by the change of address and orders were executed as usual.

Meanwhile, on De Wildt Estate the Division of Veterinary Services took over and for at least three days scores of officials worked among our poultry. Together with our staff, and in an area completely isolated from the rest of the farm, officials slaughtered and incinerated 17 000 young birds. Then all the remaining healthy hens were individually injected with a live virus vaccine, the effectiveness of which, we were informed, would only be known after two weeks. It came to light then, too, that unknown to us, one of our own farm workers, returning from leave, had brought with him a gift for his wife – three fowls. We now knew how, albeit innocently, the disease had been imported.

It took two long and anxious weeks to complete the work and the fortnight dragged by very slowly. It was heartbreaking to view daily the new heaps of dead fowls, birds that had reacted to the vaccine, and had had to be incinerated. At sunrise one day, as we sat dejectedly drinking coffee on the veranda of the house while waiting for the staff to arrive, Godfrey nodded in the direction of the hen-houses and remarked, 'Hens are cackling this morning – there must be some left.' During those weeks of hell, the fencing of the cheetah camps had been forgotten. The newly bought egg-grading machine lay unused and dusty and we doubted whether it would ever be needed on our farm; and as to any idea that we had had of altering and improving our home, well that idea simply disappeared altogether.

When the stipulated quarantine time was up and officials again conducted tests on our birds, we waited tensely for the verdict. But thankfully the latest mortalities bore no signs of Newcastle disease and so we were finally pronounced clear of restrictions. After weeks of isolation, death, incinerations, precautions and uncertainty, we felt as if spring had suddenly blossomed after a long winter. We lost no time in throwing all our energy into building up the farm again and marketing our own produce. Nor had we time to look back. Memories of the episode would remain always, but now it was seen as a thing of the past for we needed desperately to make up lost time. The new egg-grading machine was put into use and the fencing of the large 40-hectare cheetah enclosure was resumed. But shelved were the plans for a more comfortable home.

Our thoughts now turned to preparations for the formal opening of the De Wildt Cheetah Research Centre. Only three months away, the big day had been set for 16 April 1971. The Administrator of the Transvaal was to be the guest of honour; wildlife and state dignitaries had been invited and the media had been notified of the event. After the opening ceremony we planned to treat our guests to a braaivleis under the Big Tree.

The zoo had acquired nine cheetahs – three males and six females – for the initial release at De Wildt. Most of these had arrived at the zoo as cubs and we all agreed that it was important that they should remain as wild as possible if we wished them to become potential breeders. They were to be handled only when attended to by a veterinarian for medical reasons. I had visited the nine at the zoo for some weeks beforehand and had felt rather like a mother visiting her children at boarding school. The group had always spat and snarled at me, but one – later to be named 'Tana' – had been different. Raised by hand and donated to the zoo at the age of six months, she was not part of the large group and was caged separately. She had been tamed and we knew she would always remain so, for unlike other wild cats, tame cheetahs seldom turn on their keepers if well treated. I warmed immediately to Tana as she always purred loudly and rubbed her body against the bars of her zoo cage when I visited her.

On the morning of 16 April, frantic last-minute preparations were still on the go when nine large wooden crates arrived on the farm, but there was little time to examine the contents as we were expecting over 100 guests that day. Only when the speeches were over, were the nine released into the quarantine camps. Press cameras clicked enthusiastically as one by one the crates were opened and out dashed the bewildered cats. Apart from Tana, who ambled out in leisurely fashion showing great interest in the crowd of people, the cheetahs all snarled and bared their teeth at us before running off to explore their new surroundings.

There were two adjoining quarantine areas, each a hectare in size, and these were situated within the large 40-hectare enclosure. The cheetahs were to be kept in these smaller camps for a period of at least six weeks to ensure that they were not carrying any diseases or parasites. Being a true cat, the cheetah is very susceptible to feline infections, such as panleukopenia, commonly known as 'cat flu', and rhinotracheitis, or 'sniffles'. On arrival at the zoo, all the animals had been given an injection of a multiple dead vaccine, one which had no after-effect. We were instructed by the zoo veterinarians to spray the cheetahs regularly with an insecticide to eliminate common external parasites, such as mites, ticks, fleas and blood-sucking flies, and in addition to give medication for endoparasites or internal worms.

The following morning Godfrey and I were up at the crack of dawn. Wary and not yet sure of their reactions, we watched the nine cheetahs through the fence.

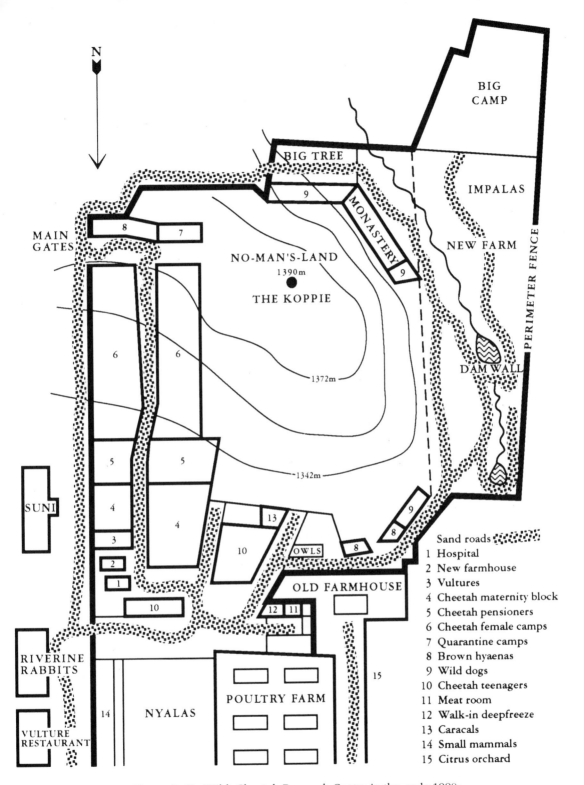

Figure 2 De Wildt Cheetah Research Centre in the early 1990s

We had expected a timid and nervous group, but found them to be just the opposite. They had obviously fully explored their quarantine camps and now they stretched their necks and with heads held high looked out through the wire fencing to the vast expanse beyond, in a manner that seemed almost arrogant to us. We were completely ignored and their aloof attitude made me realize why Joy Adamson had named her book on cheetahs *'The Spotted Sphinx'*. The general ground-colour of the fur of the cheetah ranges from a pale yellow to a golden tan broken with round black spots which occur all over the pelt, except towards the end of the long white-tipped tail where the spots merge into rings. Aptly, the word 'cheetah' is derived from the Hindu word *'chita'*, meaning 'the spotted one'. The multi-spotted coat is well designed as camouflage, especially amidst long grass and bushes. The fur is short and smooth, with longer hair over the back of the neck and shoulders. But what impressed me most were the large brown eyes under heavy brows that contributed to their expression of aloofness. Black tear-lines running from the inner eye to below the nose gave all the cheetahs a melancholy look. Their ears were small and round and overall their faces showed alertness. Supposedly the world's fastest animal, the cheetah's slim body is built for speed, with a small head, long legs, and an oval-shaped tail that acts both as a rudder and for balance. The claws, like those of a dog, are visible for they are only partially retractile and lack the protective sheaths that are found in other cat species. This unique feature enables the cheetah to grip the ground while running down its prey and probably contributes to its ability to run at incredible speed. Although classified as Felidae, the cat family, the cheetah has a number of other dog-like qualities: weak jaws and small canines, paws with hard foot-pads and an inability to climb trees easily or jump from heights.

Inquisitively, they slowly came up to the fence and I was tempted to reach out and stroke them but restrained myself remembering Frank Brand's admonition: 'Don't take chances, Ann. Remember that they are wild animals, and wild animals are always dangerous.'

As we watched them we wondered whether we would ever be able to tell them apart – they all looked so alike. Tana had been released alone in the adjoining camp as she was still too young to mix with the others in the group. As she was so much smaller we were afraid that she would be injured by the sheer weight and strength of the older animals. I was delighted with this plan for it meant that I could enter her camp and we could get to know each other whenever I had the spare time. During the first two weeks the nine settled in well and gradually we

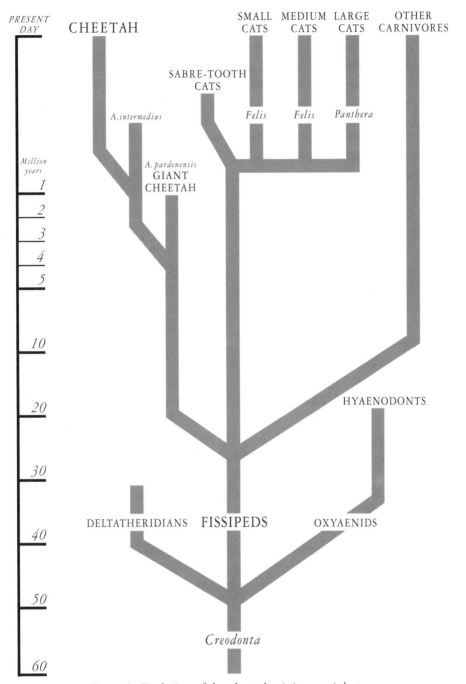

Figure 3   Evolution of the cheetah, *Acinonyx jubatus*

were able to identify each one because of both physical differences and individual manifestations of character. Two, approximately nine months old and always found close together, were obviously brother and sister and it was touching to see their care for each other. We named them 'Jack' and 'Jill'. Then there were four from the same litter: two males were more boisterous than the others and these we called 'Frik' and 'Mannetjies' after prominent rugby players, while their two sisters were 'Purry' and 'Helga'. The two other females, donated as cubs to the zoo by the farmer in the Eastern Transvaal, had already been named 'Lady' and 'Kromstert'.

The logistics involved in the feeding of all these cats could have been prob-lematical, but the zoo with great foresight had entered into formal contracts with a number of abattoirs in the Transvaal to provide us with meat. Trucks were sent regularly to these slaughterhouses to pick up carcasses which were then cut up into approximately three-kilogram pieces and delivered to us twice weekly, thus enabling each animal to be fed one portion of meat daily. Although classified as third grade, this meat is only slightly cheaper than other, superior, grades and there were the additional expenses of freezing and transporting it.

We had to feed the cheetahs every afternoon at about four o'clock. I always looked forward to this task, and although I rapidly lost my fear of the animals I continued to respect them. I devised ways of exercising the cats in the small enclosure and soon found that their greatest joy was to play with a soccer ball. Invariably, though, Purry would disrupt the game by grabbing the ball between her paws and puncturing it with her teeth. Fastening a rope to the ball and jerking it free from sharp claws in the nick of time solved the problem and so it became a ritual every afternoon, after feeding, to spend an hour or so with the cheetahs. Some would chase the ball continually while others would slyly stalk it from a distance and then pounce on it unexpectedly. Tana preferred to play with an old sack: she would snatch it from my hand and, holding a corner in her mouth and leaving the rest to drag between her legs, she would proudly make off with it as if it were her very own kill. Another of her tricks would be to ignore me completely until I passed her by; and then she would suddenly pounce on my back and purring loudly, she would give me a gentle nip on the back of my neck.

Winters at De Wildt are glorious. As a general pattern, cold sometimes frosty nights follow warm sunny days. The blue sky is never clouded, it seems, and on the odd occasion when there is rain, it quickly soaks into the parched winter earth. The long grass is tinder-dry and a constant source of concern, though every

(1) Tana and Godfrey, December 1972.

(2) The author with two-year-old Tana.

(3) *Godfrey and the author at cheetah feeding-time.*

(4) *Lady and Kromstert as cubs.*

(5) Lady taking meat from the author.

(6) Mannetjies.

(8) *The author and Dave bathing a young Erich as a preventive measure against mange.*

OPPOSITE: *(7) Johannes assisting Dave to obtain smears from a cheetah: tests later confirmed an outbreak of mange.*

(9) Dave and Denny transport a sedated Chris to the hospital for tests.

(10) *Of our first litter of cheetah cubs, Cornell was always the most eager for her mixture of egg yolk and milk.*

(11) On her return from a month-long Okavango trip, the author was welcomed boisterously.

(12) Mr John was one of four cheetah cubs found abandoned in the Timbavati Game Reserve.

(13) Dave's close encounter with the leopard released in the Mkuze Game Reserve.

attempt is made to prevent veld fires. There is a pact among farmers that each one will help the other in the event of such a fire, though once the Magaliesberg is alight it is an almost impossible task to control the blaze, especially if that fire breaks out during the night.

In the early hours of a morning in June 1971, Godfrey was woken by the ringing of the telephone. He answered it sleepily and heard the worried voice of a neighbour asking for help – the mountain was alight. The first I knew of it was when Godfrey knocked on my door and I heard his muffled voice saying, 'Ann, the mountain's burning – I'm going up. It's very cold and windy out, so don't come along. Expect me when you see me.'

'No,' I replied. 'Give me five minutes to dress and I'll be with you.'

Soon I had joined Godfrey at the garage where the cold engine of the farm truck was whining and spluttering as he impatiently tried to start it. The minutes ticked by. Time was precious as a fire waits for no one, but soon we were driving at top speed to our farm workers' houses. We woke as many sleepy people as possible, asked for their help and then we raced towards the burning mountain. When the truck could progress no further up the slope we all set out on foot, and although the sky was aglow and lit up our way, in our haste we stumbled over the loose rocks and stones. It was unnecessary to remind everyone of the unspoken rule of fire-fighting – always keep together – for it is so easy in the turmoil to be cornered between the fire and a precipitous cliff. Using wet sacks we beat the raging flames, all the while with a feeling of despair and hopelessness, for no sooner was an area under control than a smouldering tree would literally explode, shooting out lighted chips of wood that rekindled the fire in another part. Often the wind changed direction and we had to retreat from the scorching heat, coughing and choking in the billowing smoke.

Fanned by a strong wind, flames of up to seven metres high leapt into the dark sky devouring everything in their path. The deafening noise was like that of a gigantic machine – an unnatural sound in this usually peaceful setting. The only way of controlling such a devastating wave of fire is to let it burn up to the summit of the mountain – which was a stiff hour's climb away – where the vegetation is sparse and where the flames will burn themselves out. This we did, and hours later, our lungs burning from the inhalation of smoke, we were able to set off back home.

All the while I had thoughts of our new charges and how they would cope if a fire were to start in or even near their camps. It was unthinkable that they should be subjected to this danger, so the burning of firebreaks became a priority as the winter progressed. It took three weeks to cut and burn a strip about 25 metres wide inside the perimeter fence – a difficult and unpleasant task that was ably performed by a group of helpers from the zoo.

After three months we thought that it was time to introduce Tana to the others for they would all be released into the large 40-hectare area together. We did not anticipate any difficulties as the cheetahs had been in adjoining camps since their arrival and appeared to have grown quite used to one another. Early one morning we opened the gate between the two camps and encouraged the group of eight to move into Tana's territory. Lady, Kromstert and Purry ambled in first, paying no attention to Tana. But Frik, who followed them, was not so placid. He tackled Tana at once and she dropped to the ground, lying on her back and crying out. This seemed to excite the others and it was some time before we could corner Tana, lead her to safety and chase the attackers back into their old camp.

We examined Tana carefully but as we found no injuries on her we realized then that her cries had probably been of fear and not of pain. Our next step would be to introduce the group to her individually and Jack was selected first. He took to Tana straight away and we were relieved that evening to see them licking each other's faces, a sign of mutual acceptance.

Jill was the next to be chosen, though we anticipated a certain amount of jealousy with her because of her kin ties with Jack. But again there was no sign of animosity; and so, one by one, day by day, the others were slowly introduced. Frik was left until the end, and when his day came he repeated his earlier attack. This time Tana quickly scrambled up to quite a height in the nearest tree with Frik following close behind. She was unsteady because, like all cheetahs, she was not built for climbing. Afraid that she would fall or that Frik would attack her, I reacted instinctively and whacked Frik across the rump with a willowy stick that was always in my hand when I was in the cheetah camp. I waited silently for his reaction; he immediately left the tree, ignored me and also ignored Tana. Never again did Frik make an attempt to terrorize her.

We were proud that our group was now united. Tana remained friendly towards us but was initially very wary of her new companions and refused to romp with them. The tension eased after a couple of weeks though; slowly she became acceptable to the group and eventually became part of it. Not long after this, we decided that we could release the cheetahs into the 40-hectare camp. I shall never forget the obvious excitement of the nine as they strode out into the area of bush that they had been viewing for so long through the fence. Once there, they all disappeared quickly from sight and over the koppie. It was sad to think that although this was a greater freedom than ever before experienced by them, it was still not the true freedom of the natural wild. But that was something still to come.

From that day on, they all learnt to associate the sound of the farm truck with food and, at the noise of its engine as it trundled up the stony hillside track late every afternoon, they would assemble, salivating and restless at the main gates of the large enclosure. I'm afraid my family of cheetahs had never heard of Emily Post's book on etiquette and table manners. At feeding time it was a free-for-all contest when the meat was thrown from the truck into their midst. The males were the most aggressive, usually taking first pick and sometimes grabbing the meat in mid air with their front paws. However, when all the rations had been distributed, the cheetahs would run off in different directions and each one would settle down in solitude to consume its evening meal. Often two cheetahs would snatch at the same piece of meat; then, sinking down on their haunches, the meat held firmly in their jaws, neither one was prepared to give its food up. Glaring at each other and emitting snorting noises through their nostrils, they would remain passive but tense until one tried to renew its grip. Then the other, with lightning speed, would pull the meat away and run off with it in its mouth.

The jaws of the cheetah are not as powerful as those of the lion or leopard; the teeth are smaller and the jaws are controlled by well-developed muscles that enable it to hold its prey in a vice-grip manner. When eating, the cheetah either tears off small pieces of meat with its front teeth or chews on its prey in scissor-like motion with its back molars. It was interesting to watch them chew and swallow the flesh and bones of the chickens that we sometimes fed them. Somehow they managed to leave the skin with feathers in almost one piece, having consumed the rest of the bird. In the wild, cheetahs will hunt every three to four days depending on the amount of food consumed after the previous kill. However, Godfrey and I decided to feed them on a daily basis as this would enable us to keep a close watch on the cheetahs' condition, thus making it possible to spot any sign of disease or other problems at an early stage. Usually I would do this daily chore on my own, leaving Godfrey free to handle the locking up of the hen-houses.

After feeding was completed, I loved spending those few last moments before dark with my family of cats. A feeling of trust had developed between us and although they were not tame I knew they would not harm me. As cheetahs seem to enjoy a high vantage point, I invariably went to the top of the koppie from where I could see the valley stretched out before me. Frik usually took up his position on the bonnet of our jeep. Tana would lie purring at my feet and the others would flop down on the ground near us. Far in the distance the lights of nearby Brits would twinkle. Peace reigned, and sometimes I would hear a jackal call far away. Perhaps I was chasing moonbeams, perhaps I was living in a make-believe Garden of Eden, but why, oh why, I thought, could not man and nature exist always like this, in this state of quiet contentment? Surely there was room for us all?

Invariably a rough wet tongue on my hand would bring me back to reality. It would be Tana demanding my attention and I would stroke her affectionately. Her warm brown eyes always looked up at me with seeming affection and trust, and I knew then that I was beginning to experience that strong bond that can exist between wild animals and humans, once the barrier of fear has been broken. The respect and understanding that results is far greater, in my experience, than any human-domestic pet relationship. In size, Tana was the smallest of the group, and I could sense that she felt safe when I was near, as Frik and Mannetjies were inclined to bully her. These two brothers were the biggest of the group. Little did I realize at the time what an important role Frik was to play in our future breeding programme. The pair, Jack and Jill, usually lay close together, some distance from me, each licking the remains of their meal off the other's face. They had been found in an appalling condition in a remote part of the country by a Nature Conservation officer, and they accepted me only as the bearer of food. I hoped that one day I would be able to gain their trust. Lady and her sister, Kromstert, were both gentle cats and they, too, preferred each other's company at times like this.

When I shared this peaceful hour with my cheetahs I realized how fortunate I was, but wondered if I would ever really get to know them all. Would our cheetah-breeding aims and ambitions be fulfilled? Would we be able to discover why they had never bred successfully in captivity?

The serenity of these evenings was a good tonic, especially after the most frustrating of working days. I would watch the great red sun disappearing, the sky colour changing quickly from a dazzling pink to a soft pastel glow. The cheetahs seemed to have accepted me as part of the world of nature, as just another animal, and I always sensed when I was with them, a state of peace and a new-found feeling of wholesomeness. The communication call of these felids is very similar to the chirrup of a sparrow, and occasionally this bird-like whistle would disturb my thoughts and I would watch as one or other of the group responded to it. Behind me there would be a rustle of dry leaves and grass and I would turn to find one of them sniffing at the remains of a chicken carcass – that evening's meal.

With the sky turning to a brilliant deep velvet-blue I would make for home not forgetting on the way to feed Miss Kwaak-Kwaak, a pied crow with a broken wing, which had made its home in a tree just outside the main gates of the cheetah camp. I would collect scraps left over from the cheetahs' meals and place them on a roughly made platform that had been knocked together by Godfrey as a feeding table for her. At times I was concerned about her safety but she seemed quite happy and was not bothered by other crows in the area. One day, however, Miss Kwaak-Kwaak was missing and perched in her tree instead were a number of yellow-billed kites. Migrants from the north, kites visit the warm south in summer; they are brown in colour with yellow beaks, sharp dark-brown eyes and, in build, are slightly larger than a crow. They live on live prey such as small mammals, reptiles, birds and insects as well as on carrion. They are usually seen gliding up to 50 metres above the ground, their slightly forked tails, which spread out in fan-like fashion, acting as a rudder. On spotting its prey, a kite will drop down like a stone and scoop up the unsuspecting victim to eat it in flight, or if it is too heavy, on the ground. Obviously our feeding table provided the kites with an easy meal and this, judging by Miss Kwaak-Kwaak's absence, they were not prepared to share with a common crow.

I called Miss Kwaak-Kwaak by name repeatedly and eventually heard a noise in the grass some distance away. There I found her looking very ruffled and forlorn and obviously very hungry. Knowing that she had no chance against this invasion of kites I took her to our home where she took up residence in the large wild fig tree that grew outside our office window. The tree's numerous and spreading branches enabled her to hop up easily to the uppermost boughs from where she could survey her domain. Unfortunately she was a very raucous bird and soon developed a habit of calling out loudly whenever the telephone rang in the office. I often wondered what our phonecallers must have thought when they heard her continuous rasping cries in the background.

On a number of occasions we tried to banish Miss Kwaak-Kwaak to a similar tree some distance from the office, but with determined hops and loud croaks she would always return. She soon started attracting suitors and would leave a few scraps of meat on her food tray for them. Then she would watch from the branches above for any callers that came to eat. In time she became much more subdued and rarely called out and not long after this we saw that she had hooked her man and that the pair of them had started nest-building in a branch just above the corrugated-iron roof of our house. Twigs were intertwined by the two crows to form a platform; it was untidy and loosely woven, and invariably numbers of sticks would fall from it, clattering loudly onto our iron roof below. The noise was bearable when it occurred during the day, but one night Miss Kwaak-Kwaak lost her balance and dropped onto the roof with a loud thud. Then, trying to grip the iron surface with her claws, she slid down with a rasping noise onto the gutter, waking up the entire household. We decided that something would have to be done about the two crows and early the following morning Godfrey placed below their nest a large wire-mesh platform which would collect any falling sticks and, like a trapeze artist's net, would catch Miss Kwaak-Kwaak if she fell again. After raising a number of chicks, she died many years later and we had to admit that we sorely missed her noise and antics.

CHAPTER 3

The Newcomers

One day, a couple of months after the cheetahs had been released into the large enclosure, I started out as usual on my late-afternoon feeding round. After trundling up the steep dusty road, with a crate of meat bouncing on the floor at the back of the bakkie, I reached the large gates where the cheetahs assembled to wait for me to arrive with their food. But on this particular day only Tana and five others were there. After I had greeted Tana with a hug she responded with a loud purr and hopped onto the back of the bakkie to escort me, as she always did, with the others following behind the truck to the open area where I fed them. I doled out their food, all the while wondering where the other three could be. Then, when the six cheetahs were busy eating I whistled and called for the missing trio but got no response. Just as I was contemplating calling in Godfrey's assistance, there was a slight movement in some bushes below me. Hurriedly but carefully making my way down the hill over the loose stones I met Frik, Mannetjies and Helga slowly walking up to the feeding area. I was surprised to see that Frik was foaming at the mouth while the other two had a hangdog listless look.

Slowly I encouraged them by showing them their food, and they followed me up the hill. Together with Tana and the other five, who had by now finished their meal, I managed to lead them all into one of the quarantine camps. Although

Mannetjies and Helga ate a little of the chicken I gave them, Frik persistently refused to take the food and also seemed to have difficulty in drinking water.

Early the next morning a vet from the zoo arrived to help. We coaxed one of the cheetahs – which happened to be Mannetjies – into a narrow passageway which led to a crush that would confine the animal. A crush is a mobile cage just big enough to enclose an adult cheetah. It measures one metre high, one-and-a-half metres long and half a metre wide and has drop-gates, controlled by ropes, at either end. The sides are made up of horizontal and vertical bars placed approximately 15 centimetres apart which enable a vet to examine the animal and administer an injection if necessary. Spitting and snarling all the while, the aggressive Mannetjies was quickly sedated. It was a strange sensation to be able to stroke him, now that he was rendered immobile and I was able to remove ticks from his body and to feel his smooth fur. After a thorough examination the vet found that his whole mouth – his palate, throat and tongue – appeared very inflamed and was covered with small white blisters. Blood samples and swabs from the infected mouth were taken, and sent to the research institute at Onderstepoort for analysis, and the sleeping animal was given an antibiotic injection. His temperature was 41 degrees centigrade – three degrees above normal.

After much persuasion Frik and Helga also found themselves in the crush and although the vet did not sedate them, they too received an injection of antibiotics. Over the next few days, as one by one the other animals went down with the same infection, they were treated. Godfrey and I were naturally very worried about our cheetah family and visited them as often as possible during the day. Usually we found them lying under the trees, listless and subdued and showing no interest in their food. As I had done earlier with Frik, we tried to make eating easier for them by removing the meat from the chickens and then cutting it into small portions; we also gave them pieces of wild game meat. But all was to no avail. One comforting fact was that, although they achieved it with some difficulty, they were able to drink water.

In time the report came back from Onderstepoort stating that the cheetahs had indeed contracted a bacterial infection, and we were advised to administer antibiotics daily for five days. These days passed slowly and to begin with the animals showed very little response to the drugs. We tried to minimize the stress that they were subjected to when they were chased daily into the crush for their antibiotic injections, for we had noticed that the least amount of exertion left them listless and panting heavily. Tana was the worst affected and I spent my free time sitting next to her and stroking her gently. I realized then how much a part of my life she had become – and also what it would mean to me if I were to lose her. She always responded to my touch with a throaty purr and then she would rest her head on my lap. At the same time, uncharacteristically, the others made little attempt to move away from me if I got close to them. At last, very slowly, we observed signs of recovery. Their eyes became brighter and more alert. They seemed to have a little more energy, and with joy we soon noticed that they had started to show more interest in their food. After two weeks all were fighting fit and could be released once more into the large enclosure.

On several occasions as we went on our daily feeding round, Godfrey and I had caught glimpses of a foreign animal in the large camp. Its coat was a golden brown in colour and it seemed to be the height of a medium-sized dog, but it darted into the grass and bushes with such speed that it was impossible for us to identify it. We thought that it could be a duiker or a steenbok, or perhaps even a jackal. But when one day we found cat spoor clearly marked in a patch of fine sand near the cheetah quarantine enclosures, we decided to bait the animal with meat and in so doing entice it into one of the small camps. With a large chunk of topside tied to a cable which in turn was attached to a spring in the gate, our trap was set. Early the following morning, we went up to the enclosures. To our delight – even from a distance – we could see that the gate was firmly closed and we knew then that we had caught our animal. We hurried up to the camp to find to our amazement that we had caught not an antelope but a large male caracal – or 'rooikat', as this wild species of cat is commonly called.

'Oh, Godfrey! What a beautiful cat!' I exclaimed excitedly.

'Unbelievably handsome,' he replied. 'But what a shame – we must have unknowingly fenced it in.'

My excitement soon turned to pity as I watched the captured animal race back and forth in the enclosure, hissing and snarling continuously.

'Shouldn't we release it further up on the mountain?' I asked. 'It has obviously lived in these parts all its life and it should know the area well and can clearly fend for itself.'

'Let's first give it a little thought,' Godfrey replied wisely. 'We'd better discuss it with our neighbours – it would be sad if it were to be shot.'

The caracal is found in savanna areas throughout Africa. Solitary and nocturnal in habit, its natural prey is small antelope, dassies, hares and birds, but to humans it is seen as a wanton destroyer of livestock for it attacks and kills large numbers of sheep, goats or poultry and eats only parts of the carcasses. Powerfully built, it is cunning, agile, and one of the fastest of the smaller African cats, attacking without hesitation when cornered, or should its young be threatened. Its cold and darting pale green eyes and unusual ears attracted our attention immediately. The backs of its ears were covered with fine short black fur which tapered off in long tufts at the tips – hence the name 'caracal' which is a Turkish word meaning 'black-eared'. The robust body of the cat is covered with a smooth, soft but thick reddish-brown coat, changing in colour to an almost white underbelly. Its muscular hindlimbs are slightly longer than the forelimbs and the tail is strangely short. Being a good climber and agile jumper, it often surprises flocks of pigeons or doves and, with unbelievably fast forepaws, swats and kills its prey.

The brown of its body seemed to blend with the dry earth as, wild and terror-stricken, it rushed this way and that. We decided to leave it alone until it had calmed down and we could decide on its future.

Although we were very tempted to release the caracal into the wild we felt that there its chances of survival were indeed limited. For some time it had apparently been feeding off the occasional pieces of chicken left by the cheetahs and, having thus acquired a taste for this meat, it would, if freed, probably raid chicken-houses in the neighbourhood and in time have to be destroyed. The zoo, however, solved our problem for it turned out to have a tame hand-reared female of the same species – 'Snippie' by name – and she was duly sent to us as a companion for our caracal. By now our captured rooikat had been named 'Oubaas'. He was dignified, definitely the boss and to us he seemed to have an expression of aged wisdom. Snippie was much smaller than Oubaas who seemed aware of the superiority that size gave him. The two cats tolerated each other and accepted captive conditions without any further ado. Although caracals breed easily, whether in the wild or in captivity, we could not merely assume that these two would produce offspring; nevertheless we still looked forward to the possibility of having caracal kittens on De Wildt Estate one day.

Our neighbours have always been and still are most interested in the cheetah project. Often we are asked if we can be accompanied on our early evening feeding trips during the weekends when farming folk have a little more spare time than usual. When one lives in the country there is a greater bond of friendship between neighbours than is normally found in cities – perhaps because we are fairly isolated and dependent upon one another for company and in times of need. One Saturday afternoon while Godfrey and I were out on our feeding round, Boet and Rita Schalkwyk and their young nine-year-old son, Andries, found no one at home when they called and so they decided to look for us. A large 'no entry' sign had been on the main gates ever since the cheetahs arrived. Not knowing that the animals had changed their quarters and were now wandering free in the large enclosure, our friends ignored the sign believing that the cheetahs were all still safe inside the small quarantine camps. In the distance Godfrey and I heard the sound of a car engine. It stopped, and with dismay we realized that its occupants were now walking through the large camp. Our fears were confirmed when Lady suddenly left her food and headed off in the direction of the gate. Godfrey shouted to our guests, trying in as few words as possible to warn them of the danger. Perhaps they could not distinguish what he was saying for they appeared to be unconcerned and merely continued to walk towards us.

By now the other cheetahs had become inquisitive as well and had disappeared in the direction that Lady had taken. In the camp, Boet, Rita and Andries were suddenly confronted by the group of cheetahs and all three instantly froze in their tracks. Lady, backed by her mates, ignored the adults and with eyes set on young Andries, crept slowly, head close to the ground, past his parents to the now immobile, rigid boy. Godfrey and I dashed towards the group in time to see Lady gently placing her front paws on Andries's shoulders. We held our breath and watched the boy who, fortunately, did not panic and remained motionless. There was nothing savage in Lady's action and her claws did not even penetrate his shirt; there was merely a look of complete surprise on her face as if she were thinking, 'Look at what I've caught!'

Just as gently and as slowly as she had placed her paws on Andries's shoulders, Godfrey lifted Lady from the boy. As an onlooker, I felt that I was watching a slow-motion sequence from a dream. It took only a few minutes before I was able to shepherd three shocked visitors from the camp while Godfrey remained behind with the cheetahs. Since that episode, needless to say, whenever we enter a camp the gates are firmly locked behind us and adherence to the 'no entry' sign is strictly enforced. We have noticed that all the cheetahs continue to display an interest in children, especially the smaller ones – perhaps because of their size they feel that they can handle them so much more easily than adult humans.

Towards the end of 1972 the females showed signs of coming into season. Typically cat-like, they rolled on the ground and presented themselves to the males. Jack was sexually the most active of the group but found life not as exciting as it could have been as Frik was the dominant male and was constantly putting him in his place. On one occasion, while Frik was busy eating his meat, Jack realized that he was alone with Lady and made advances to her. She rejected him sharply with a fast swipe of the paw and her growling quickly brought Frik to the scene. But there was no animosity this time, for Jack simply turned his back and strolled off with an air of complete indifference. Mannetjies had no standing at all in the group and was always aggressive towards Godfrey and me. Never quite sure of how he would react in new situations, I suggested to Godfrey that we should place him in one of the small enclosures during the breeding season. But Godfrey felt that we should give him a chance as he seemed to add life to the party, and so Mannetjies remained with the others.

One warm late afternoon I was alone in quiet and drowsy contentment while waiting for the animals to finish their meal so that I could gather up the scraps for the crows to feed on. Tana, who was a small eater and always the first to finish, lay as usual at my feet licking her front paws and purring. The only other sound in the still air was that of the rest of the group breaking and crunching chicken bones. I leant back against the side of the truck enjoying the peace of the late afternoon. Suddenly, without warning, panic broke out among the cheetahs. What caused it I will never know, but with lightning speed Mannetjies came hurtling towards me, knocking me clean off my feet. Both on the ground and completely bewildered, we slowly disentangled ourselves and Mannetjies and I for a moment found ourselves looking at one another eye to eye. In that brief second an unwritten truce seemed to be declared, for after that Mannetjies showed little aggression towards me. I was never to discover the cause of the panic – probably it was triggered by some sound inaudible to me – but through the experience I learnt of yet another side to the highly strung nature of the cheetah.

In spite of what had seemed a promising beginning, nothing further materialized during the mating season and we began to have grave doubts as to whether we would ever be successful. Something was wrong in the social set-up, but what was it? Something was missing, it seemed, but what was the link? Was the family possibly too happily close-knit? Could this be nature's way of preventing inbreeding?

At the outset of the project I had tried to obtain information on any research or work that had been done on the breeding of cheetahs in captivity, but I discovered that very little had been written. However I did come into contact with Nan Wrogemann who was collecting all available literature for a book which was subsequently published as *Cheetah Under the Sun*. As we had similar interests we became good friends and Nan helped me to obtain most of the available papers that related to captive breeding. According to this information, in the 18-year period from 1956 to 1974, 29 females produced only 140 cubs in zoos and game-parks throughout the world – a birth-count that is considered very low indeed. To make matters worse, the survival rate of the cubs to four months of age was a mere 57 per cent.* These figures indicated that very little had ever been achieved in the field of cheetah-breeding in captivity, creating an even greater – yet challenging – task for us. In an effort to find an answer, Godfrey and I considered the close-knit character of our group. We suggested to Frank that further cheetahs should be brought in to break up the familiarity existing between our nine and to introduce new genetic strains. He agreed with this proposal.

Some months later, routine was disrupted early one morning with a telephone call. We were told that 16 cheetahs originally caught in the wilds of Namibia were being sent to us by the zoo. Confiscated by Nature Conservation officials, these illegally held animals had been found in a state of poor health on a farm in the Eastern Transvaal. Although it was then mid winter we were not perturbed about the translocation of these cats as our own climate, being sub-tropical, is similar to that of the Eastern Transvaal.

'The animals are on their way to De Wildt at this very moment,' said the caller from the zoo. 'And like the last group, they must be kept in quarantine for at least six weeks. Can you foresee any difficulties?'

Fortunately our quarantine camps were empty so it was only a matter of arranging extra food and water for the 16 newcomers.

'No, not at all,' we replied. 'We'll cope!'

We quickly set about preparing for the new arrivals. Our enthusiasm soon turned to dismay: what a shock we got when the crates carrying the 16 were opened, for the cheetahs were all in a most pitiful condition. Bones protruded through scratched and dull fur and open septic wounds on all of them caused us to stare at them in horror. They were the complete antithesis of our original nine sleek and healthy animals.

Before releasing the new group from their crates, each one was inoculated against infectious feline enteritis, and penicillin was given for the infected wounds. The group consisted of 14 males and two females, their ages ranging from two to 10 years. They were a wild crowd, seemingly frightened and under-standably so when one saw their sorry state. One of them, Mof – named after a well-known rugby player – was extremely aggressive, this behaviour being unusual for a cheetah. Mannetjies's aggression was mild compared with that of this young cat. With head down, eyes glaring upwards to show the lower whites, and tail thrust forward between the hindlegs, Mof was tense and hostile.

It did not take us long to realize that the sexes would have to be separated and the only two females of the group, Nan and Judy, were drawn off into an adjoining enclosure. Two of the males, Anton and Fanus, had festering sores that needed constant attention and we wondered at times whether they would survive. But with regular feeding and medical care they and the rest of the group gradually improved. They began to put on weight and their bony outlines became more elegant and rounded. As their wounds healed, so sleek and shiny coats emerged. But one thing never changed: Mof always appeared to hate us.

Meanwhile, Snippie the caracal was due to have kittens any day. Late one afternoon, at feeding time, I found her in labour and stayed with her until dark. Although she was having contractions, nothing further happened. I was up early the following morning and hoped that I would see the new caracal kittens for the first time. But when I reached the camp there was no sign of them, and Snippie, furthermore, appeared to be in great pain. Strangely, she accepted me in her enclosure and allowed me to stroke her. I stayed with her as long as I could but eventually had to leave her to attend to the busy month-end farm activities. Returning to her at midday, I found Snippie walking excitedly around the camp with the hindleg of a kitten protruding from her vagina. As soon as she saw me she came forward and rubbed against my legs as if asking for help. Slowly and gently I managed to release the kitten – but it was dead, still-born with a malformed hindleg. I left Snippie in her grass hut, sleeping heavily. Later in the day another kitten was born – but it too was dead.

There was no time to think about these losses, for within days a black-backed jackal pup arrived from the Western Transvaal to join us. She was petite and nimble on her feet and we named her 'Marie Antoinette'. At about the same time we were presented with two German Shepherd pups, Bismarck and Kaiser, and the three of them played together all day. Light-footed Marie teased the two seemingly clumsy dogs unmercifully, darting deftly between their legs. Chasing cats was another of the trio's favourite pastimes, but once a felid had sought refuge in the nearest tree, Marie Antoinette lost interest and left the two dogs barking at the base. We admonished Bismarck and Kaiser constantly and found out only later that Marie Antoinette was in fact the instigator of the trouble. The antics of the three gave Godfrey and me many a good laugh.

Within a few months the new arrivals were all in good condition and, having been successfully dewormed as well as sprayed with insecticide to combat an unusual type of blood-sucking fly (*Hypobosca*) that they had brought with them to the farm, we decided that it was time to introduce them to our original family of nine. First we brought our six females into a camp adjoining the 14 new males. This done, it took no time at all for the foreign males to make their way to the fence dividing the two camps and to show great interest in the goings-on next door. The females, on the other hand, appeared to be nervous and were sub-missive, with the exception of Tana and Lady who acted aggressively towards their inquisitive neighbours. Next we placed our three males in an enclosure adjoining the two female 'Namibians', as we named the group. Here, the reaction was exactly the opposite, for Frik, Mannetjies and Jack remained distant, refusing to have anything to do with the newcomers. That seemed fairly promising – aloofness was easier to handle than aggressiveness.

Two weeks later we opened the interleading gate and let our three males into the camp of the two Namibians. The males were the first to react, uttering a low guttural sound followed by a chirr and then a purr. They immediately started to 'read the newspaper', as Godfrey called the phenomenon: crouching down on their forelegs, they scanned and sniffed at the ground, especially close to tree trunks, occasionally lifting their heads with a strange almost puzzled grimace on their faces. They were smelling the urine of the females to establish whether they were in season or not. This behaviour lasted for at least half an hour during which the males marked their new territory by urinating against trees and large rocks. Only then did they begin to acknowledge that they had company in their camp. While the sniffing antics were being performed, the two females had been watching every move of the males from what seemed to be a selected hiding spot some distance away. Once the males sensed that they were there, a certain amount of chasing followed. But the females were not in season, we concluded, for they were dealt a number of quick blows by the males to which they retaliated by lying on their backs and crying out submissively. Having established their dominance, the males then lost interest and left them alone. The move itself, however, appeared to have been successful.

Bearing in mind that our original males were not nearly as aggressive as the newcomers, we decided to introduce only five of the 14 foreign males into the female enclosure. The same procedure followed once these males were inside the camp: they, too, smelt the ground and urinated. But this behaviour came to a halt when one of the males suddenly attacked Kromstert. Lying on her back as the other females had done, she was set upon by first one and then the remaining four. It was some time before we managed to separate them, though apart from a few scratches the bullied female was unharmed. It was only a matter of days before the group had settled down reasonably well. Lady and Tana, however, remained unresponsive and kept their distance as if they did not seem to care for much communication with the new males.

As this group appeared to remain fairly compatible, we decided that we could release them into the large enclosure. On the day chosen, Godfrey and I followed them into the camp until they split up and headed off in different directions. The bush was thick and the grass long and they soon disappeared from view. Tana alone held back, almost asking me to accompany her. I dearly wished that I could have gone with her, but realized that my presence would have affected her relationship with the others. All that I could do was to hope that the next day when I returned at feeding time, our family and the newcomers would be waiting at the gates for their food.

* See Wrogemann, 1975. *Cheetah Under the Sun*. Johannesburg: McGraw-Hill, pp. 136-138.

CHAPTER 4

Tana

*G*odfrey and I were pleased that the introduction of the two groups of cheetahs to each other appeared to have gone well. We knew that the newcomers had been born in the wild, that they were from different families and that most of them were mature animals. Among the males there must surely have been some that had fathered cubs in the wild. Perhaps now, with this new blood, the pieces of the breeding jigsaw would fall into place. Although the 40-hectare enclosure gave them sufficient space to move about in, and to hide away in if fighting occurred, I was still uneasy about Tana. Even though she was mature, she was of small build, gentle and not a fighter. The fact that she was tame must have contributed to her passive nature. My uneasiness increased as the day of their release drew on and in the early afternoon, as if sensing my anxiety, Godfrey approached me saying, 'Ann, I think we should go and feed the cheetahs now. It'll give us sufficient time to check on the animals and see if all is well.'

'Good idea,' I replied with relief. 'I feel strangely unhappy about them. And in any case the camp is very big – if they're not at the feeding spot when we get there, we'll need the time to find them.'

One of the farm workers, Joel Tollo, had by now been transferred from the chicken farm to work full time on the cheetah section. He showed no fear of the cats and liked the responsibility. After helping us to load the animals' meat onto the back of the bakkie, he joined us in making our way up the road to the top of the hill. It was a warm afternoon, for spring had arrived. I noticed that the leafless trees had suddenly burst forth with new growth, producing a kaleidoscope of different shades of green. We had had some light showers during the past few weeks and the grass of the burnt firebreak strips was pushing up vigorously, turning the blackened veld into a sward of bright green. It was a beautiful time of the year, I thought, with winter behind us and the long summer days round the corner.

Nearing the crest of the hill my thoughts were of Tana. A tremendous bond had developed between us. It is difficult to describe what it is like to experience trust and to receive such warmth and gentleness from a wild creature. I sensed a kind of dependence that the other cheetahs had on Godfrey and me, but with Tana it was something different and very special. Tana's attachment to me went beyond the fact that I provided her with food.

As we approached the gates, my heart seemed to stop beating. My earlier fears were not unfounded, for Tana and three of the new males were missing. After quickly feeding the animals that had gathered there, the three of us, Godfrey, Joel and I, separated, each going in a different direction, searching, whistling and calling. Half an hour later we met again, each with nothing to report. Having covered the southern slope of the koppie – for that was the direction in which the cheetahs had headed earlier – we now set about tackling the northern side. While moving down close towards the house I thought I saw a movement in the bush and grass, but I couldn't be certain. Suddenly I heard sounds of what was obviously a number of cheetahs fighting – the shrill cry of fear of a female sounding high above the other noises.

Clambering down and stumbling over the stones, I could not seem to get to the source of the discord fast enough. Eventually I came across Tana lying on the ground surrounded by the three males. On her face there was a look of wide-eyed terror, an expression of intense fear which I had never seen before; as I approached the group she turned towards me. She was positioned in a corner of the camp but, armed with a small branch that I had hurriedly broken from a nearby tree, I managed to wedge my way between her and the angry males, continuously waving the branch as I moved. As the cheetah is really non-aggressive by nature, it was reasonably easy to keep the three at bay for a while. I knew that soon Godfrey and Joel would come looking for me. Tana remained motionless, now keeping her eyes on her attackers, and I silently prayed that she was not badly hurt. I saw nasty lacerations on her hindquarters and tooth-marks on her legs.

Looking at her lying there, I could see why it had been easy for the males to pick on her: being the smallest of the group she was no match for them.

Godfrey and Joel had heard the noise of the fighting and it was not long before they joined me. Quickly we assessed the situation and acted: Godfrey sped off to the poultry farm to get help, leaving Joel and me to guard her. We planned to erect a fence quickly, that would temporarily protect Tana for the night.

Dark storm clouds were by this time gathering in the west, blanketing off the sun and causing the light to fade swiftly. Joel kept a watchful eye on the three growling males while I sat with Tana, stroking and comforting her. Although she was badly bitten, her back was fortunately not injured. She relaxed slightly at my touch, but her eyes remained firmly focused on her aggressors. It started to rain, not heavily, but enough to soak us, and for once in my farming life I wished it would stop. Where was Godfrey? Why was he taking so long? The storm moved closer; a cold wind gusted, lightning lit up the sky and thunder rumbled in the distance. Then Joel and I heard voices and Godfrey appeared followed by helpers carrying torches, netting wire and canvas. By now it was all but dark and the storm was about to break above us. With frenzied haste, a small enclosure was erected round Tana and with Godfrey's usual thoroughness he ensured that not even a rat, let alone a cheetah, would have been able to enter it and disturb the injured Tana that night. Just as the canvas was being fastened over the top, the heavens opened and the rain poured down. We left Tana with a freshly slaughtered chicken and the knowledge that at least she would be quite safe for the night.

At first light the next morning I made haste to attend to Tana. Her wounds were ugly and obviously painful, but she purred and allowed me to handle her. After discussion with the zoo it was decided that the best place for her would be their hospital where she could be treated intensively. She would probably need stitches and almost certainly medication to counteract infection. With difficulty, for she clearly moved with great effort, we gradually coaxed Tana into a crate which was loaded onto a truck. As the vehicle drove off, we turned our attention to the three trouble-makers. They had remained close to Tana's temporary enclosure and were not difficult to find. Still aggressive and threatening, they approached us slowly but dribbled at the mouth when they saw the chickens that we had with us. We tossed the birds to them without hesitation and each one grabbed a portion and disappeared into the bush to eat it.

It was nearing midday and following the previous night's thunderstorm it was both extremely hot and humid. Wearily we returned to the egg-room to dispatch rather belated orders. As always, the business kept us busy but our thoughts were not on our work and we had to admit that the cheetahs had become very much a part of our lives. The staff working in the egg-room were competent and reliable so Godfrey, Joel and I felt confident that we could escape earlier than usual to feed our felids.

We set out in the bakkie with the daily rations, each one silently wondering what we would encounter. Nearing the gates to the large enclosure, Godfrey muttered, 'Damn – no Lady or Kromstert!'

'Damn!' I echoed more feelingly.

Joel frowned, his eyes scanning the camp. After feeding the group that had gathered, the three of us separated once again to search for the two missing sisters. As we parted, Godfrey said, 'We'd better concentrate on the thickly wooded area over on the west side – I don't know why, but I feel sure they're there.'

He was right, for a little while later I heard Godfrey calling us. Under a bush, barely visible, were the two females. Lady, we discovered, had tooth-marks on her haunches – not serious ones – which she had obviously been licking. Kromstert was unharmed and had clearly spent the night at her sister's side. Marking the spot mentally, we left them with food and water. The incident was not serious enough, we felt, to lock them up in the quarantine camp. We knew that the cheetahs just had to be given time to sort out for themselves their order of dominance.

'How is Tana?' I enquired of the zoo the next day.

'Her wounds have been attended to, but she won't eat,' came the reply.

'I've an idea,' I said. 'I'll send her some chickens tomorrow – she always prefers poultry to meat. I'm tempted to come myself and see her, but think it'll only upset her,' I added.

'That's fine,' came the response. 'Do anything that'll help.'

The following day, while on a business trip to Pretoria, Godfrey made time to visit Tana.

'She's in a bad way,' was the news he gave me over the telephone from Pretoria.

My throat tightened. 'Nothing must happen to Tana,' I said.

'The zoo vet thinks that your presence will induce her to eat,' Godfrey said. 'I agree that you should come and see her. Tana seems so bewildered by the strange keepers around her.'

'I'll leave immediately,' I replied. 'Do you think I could stay in the zoo hospital?' I asked, desperately keen to do anything that would assist Tana.

'That won't be necessary,' Godfrey answered. 'I've spoken to Sue Hart and she's offered to keep Tana at her house and you can stay there – isn't that the answer?'

And the answer it certainly was. Sue Hart, a well-known wildlife enthusiast and veterinarian, who happened to be on leave at the time, was free to give Tana constant medical attention. She lived in a suburb of Pretoria and worked during the day at the Mammal Research Institute of the University of Pretoria. I would be able to stay in her home and help Sue in any way that I could.

I was shocked to see Tana when I arrived at Sue's house. She had become extremely thin and from her position as she lay on a rug on the ground she was unable to lift her head. She barely opened her eyes when I bent down and touched her. A hard lump constricted my throat as she purred gently when I called her name softly. Sue, with wisdom and sympathy in her eyes, took my hand and said simply, 'Think positively, Ann.'

I moved as if in a dream. This was not the Tana that had romped with me on the koppie, flushing out the guinea-fowl or francolin that dared to settle near us; or the Tana that after feeding had lain at my feet in the late afternoon, purring and sharing that peaceful hour with me. Surely I would wake and find my fears were unfounded. But Sue's voice brought me back to the present. She instructed me to feed the languishing cat an egg flip – a mixture of raw egg yolk and milk – and little by little with the aid of a spoon the fluid ran slowly into her mouth. I spoke to her continuously as the egg was swallowed with great effort on Tana's part.

'She's very anaemic,' remarked Sue. 'I wonder whether she's got biliary as well.'

'Is there a test you can do to find out?' I asked. I was eager for any sort of development that would help to ascertain the cause of Tana's rapidly deteriorating condition.

'I don't have the necessary equipment here or at my office,' answered Sue, 'but I can phone a colleague, Dave Meltzer. He lives close by.'

Dave Meltzer was not on duty that day but his partner, John Norman, was and he willingly offered his help. From blood smears taken from Tana he found no biliary, but established that she had a very low level of red blood cells. A drip containing intravenous fluid was set up immediately while Sue again dressed the septic wounds which had begun to fester terribly. As I helped her to turn the animal over, I rested Tana's head in my left hand and my index finger somehow managed to stray into her open mouth. At that instant Tana must have experienced great pain, for her jaw clamped down on my finger, completely severing the top portion.

At the same moment she opened her eyes and stared at me with a look of what seemed to me great sadness and I wished I could know what thoughts were going through her mind.

Dear Sue now had two patients on her hands, for she had to get me to hospital immediately. My finger was extremely painful and I felt faint but tried hard not to show my feelings. Sue, however, sensed my suffering and before we left for the hospital she contacted and advised the surgeon on duty that we were on our way. On our arrival the doctor and staff were waiting and before they wheeled me into the operating theatre Sue handed them a bundle of gauze – within it was the top of my finger that she, without my knowledge, had retrieved. Once I was in the theatre, the surgeon gave me a local anaesthetic and after thoroughly disinfecting the injury he proceeded to sew on the severed tip of my finger. He chatted to me all the time in order to take my thoughts off the operation, and the attending nursing sisters constantly asked me questions about my involvement with the cheetahs.

It was after one o'clock when Sue and I returned to her home to find Tana asleep. During the rest of that night Sue got little assistance from me and had to do the changing of drips and checking up on the cheetah's condition unaided.

In the morning Tana was slightly better. Dave Meltzer was now back at his surgery and he popped in every now and then to assist us. We soon found that Tana could not have been in better hands. Although she was still very weak, Dave thought that she could return to the farm the next day, and he together with his wife, Miriam, offered to transport her in the back of his station-wagon. Godfrey had prepared a grass bed in anticipation of the patient's arrival – on the floor of our sitting-room. This room had been specially selected as Godfrey thought the view of the garden through the large glass doors would be more akin to Tana's natural surroundings. Also, the doors would let in a welcome breeze. Dave and Miriam stayed to see Tana settled down, but soon after they had left, she took a turn for the worse. A frantic telephone call brought Dave back over the 50 kilometres that he had just travelled. He set up a drip immediately and only when Tana was sleeping peacefully and seemed much more relaxed did Dave leave.

That night I dozed fitfully next to her, changing the drip bottle when necessary. My finger throbbed painfully but it seemed unimportant at the time. While I comforted and stroked Tana, she as always purred at my touch. In the early hours of the morning I was awakened by a sharp cry and peering at Tana closely, I saw an expression on her face that I had learnt to recognize as one reflecting great pain and fear. I knelt beside her and stroked her gently and in response she lifted her head slightly and gazed at me with her large brown eyes. Her breathing was slow and laboured and her limbs cold. Suddenly she froze, her head dropped back . . . and she was gone. Then I noticed that tears rolled down her long cheetah tear-lines.

I could simply not believe that the moment I had been dreading had arrived. I called her name softly, my tears mingling with hers. There was no response. The miracle of reawakening that one always demands of death did not occur.

If only we could have been given the chance to relive this chapter in our lives, we surely would have been able to act differently. We questioned; we if-onlyed; we mourned. Godfrey and I were shattered by the loss of one so trusting and so close to us. Afterwards we found that most of the pleasure of feeding the other cheetahs each afternoon had gone, for there was no Tana. There was no Tana to meet us at the gate as we approached in the truck. There was no Tana to ride on the truck with us to the feeding place. There was no Tana to purr and lick our hands when she had finished her meal. All these things she had done of her own accord – they were part of her character and these manifestations of her affection had endeared her to us. This had been a wild animal that had sensed love and protection and, as we saw it, she had returned it in her own animal way.

But, as always, time's healing eased the pain and slowly we picked up the broken pieces and started again. We began to look forward to feeding the cheetahs in the afternoon, but we knew that for a long while it would never be quite the same. My finger healed too, although the graft was not a success. Its now shortened length is a permanent reminder of a very sad episode in my life.

CHAPTER 5

The Challenge

*I*t was many months before our enthusiasm returned. But looking back on it all now, I realize that there was a positive side to Tana's death, for through it we gained the friendship and help of Dave Meltzer, a man who was to play a major role in the breeding of cheetahs on our farm. He showed tremendous interest in the project right from the start and although he had a very busy veterinary practice to attend to, he offered to help us in an honorary capacity. Often when we needed assistance he would arrive at De Wildt late in the evening having first attended to all his cases in his consulting rooms. The three of us now agreed that the next plan of action in our attempts to breed cheetahs was to choose what we thought were the most suitable males from the new group. We selected five that we thought were aggressive but did not seem inclined to attack the females wantonly on sight. By introducing them to our five females one at a time at intervals of two days, we managed to get them to settle down together comparatively peacefully – apart from the few blows intended purely to show who was the Boss. The groups tolerated each other well, the sexes keeping very much to themselves.

One afternoon, just as we were beginning to feel happy about our grouping, Godfrey and Joel went to feed the cheetahs. On reaching the gates of their camp they heard familiar high-pitched submissive calls together with aggressive deep growls. Only four females, they soon discovered, were waiting for their food. After quickly feeding them, Godfrey and Joel went immediately to the spot from which the battle sounds had come and found an injured Lady lying on the ground surrounded by the five selected males. Keeping the aggressors at bay, Godfrey sent Joel down to the farm to call me and also to summon additional helpers. We abandoned chores in the egg-room and minutes later found Godfrey and the problem animals. While he continued to guard Lady, the rest of us circled off the attackers and manoeuvred them into the nearby quarantine camp. With the bullies out of the way, we coaxed Lady into a smaller, quickly erected enclosure. She had been badly mauled – we saw several deep gashes in her hindquarters – but thinking that the journey to Pretoria would upset her further, we decided instead to ask Dave to call at the farm as soon as he could.

Lady, who was not tame but had always been calm and quiet in character, was now showing signs of pain and shock. When Dave arrived late that evening she, with lightning speed, lashed out with her front paws whenever he got close to her. Eventually, however, we were able to pin her down in a portable crush where, after sedating her, Dave cleaned and treated all her wounds, stitching where necessary and bandaging both backlegs which were badly lacerated. Finally, after an injection of penicillin, Lady was left in a small fenced-off area for the night, with a good chunk of meat and a bowl of water next to her.

For the following two weeks Dave managed to fit a daily visit to Lady into his tight programme. When the bandages were removed after the first week, he found that her legs were healing nicely and only one limb had to be rebandaged. She ate well, and it was not long before she had recovered completely and could return to the female group. I remembered Tana and breathed a quiet sigh of relief.

We once again kept the sexes apart until it became apparent that some of the females were coming into season: they rolled on the ground and with a bird-like chirp called submissively to the very interested males on the other side of the fence. Among the Namibian group, Al, a dark-coloured male, showed interest in Nan – one of the two females we had separated from the group soon after their arrival. We let him into her camp and he covered her immediately. After copulation, which lasted for a few minutes, there was a great deal of rolling on the ground on Nan's part with Al sitting close by and growling softly. Godfrey and I were thrilled. At last, it seemed, we had achieved a small breakthrough – and all this had happened on Christmas Day!

Al's interest in his new-found mate lasted another day and then they went their separate ways. In the wild, the cheetah female is generally solitary with the male making his appearance only at mating time. She is left alone to fend for herself during pregnancy, and later for her cubs when they are born – a hard task, considering that she has to run down her prey, and her kill could be located some distance from her youngsters. While she is out on the hunt the cubs are left hidden in the tall grass or thick bush – easy quarry for other predators such as lions, hyaenas, jackals, wild cats and even birds of prey. To add to the hardships of the mother, lions and leopards are known to rob cheetahs of their kill, for, being so timid, the cats will choose to give way rather than fight. Hence, when cheetahs eat they are always on the alert and gorge themselves quickly, seldom returning to the carcass.

After a great deal of observation and prompted by recent setbacks, we decided to place only one male with a female in season. If we could observe mating taking place, we would be able to calculate the approximate date of birth of the cubs and at the same time establish the parentage of the youngsters. We had learnt to recognize each of the 31 cheetahs individually from their facial markings and this enabled us to pair off animals that were not related. Few females showed obvious signs of oestrus (or coming into season), but we did observe that if a female was at the peak of her cycle, she would accept any available and interested male. According to research work done in the field, the male is thought to be territorial, continually marking off his area with urine or scent marks, while the female is more free-roaming, seeking out a mate only when she is in season. In the wild, cheetahs are known to breed throughout the year, but we found the females to be more receptive during the warmer months, from November to March. During this period we had observed Al mating with Nan and later with Judy and we felt confident that they were both pregnant. They were accordingly placed in camps of their own, and we marked off on the calendar the 93 days reported to be their gestation period. The two enclosures were placed out of bounds and were visited only in the afternoons at feeding time or when Dave called as he regularly did to check on their condition. We happily watched their bodies develop, their abdomens swelling noticeably and the distended mammary glands forming a definite milk-line.

Then, one afternoon, five days before she was due to give birth, Nan could not be found. We decided not to search for her, but rather to wait until Dave called the following morning. However, to our surprise, the next day Nan was waiting at the gate when we approached the camp and hungrily took the dead chicken we had brought with us.

It was an anxious time for us all. Had Nan aborted? Were the cubs dead within her? As she still looked healthy and pregnant, all we could do was to wait patiently for further developments. Then the weather changed suddenly. It started to drizzle and became very cold: by now it was late April and winter was approaching. Nan was beyond her 93-day gestation period and Judy was very close to it but both continued to come happily for their food every afternoon. The time passed and with every day we marked off on the calendar, our faces grew longer. When both females were well overdue, slowly the size of their abdomens began to subside and Dave told us that they must have had false or pseudo-pregnancies. It was a great disappointment, but it made us more determined than ever to continue with our project.

After the setback of Nan's and Judy's 'pregnancies', Dave, Godfrey and I decided that allowing the cheetahs to wander freely in the 40-hectare enclosure had not been a success as far as breeding was concerned. We felt that smaller camps and greater control were needed. However, we knew that our attempts of the past two years had not been wasted as in this time we had learnt how to treat, handle and care for the big cats. Our efforts so far had been on a trial-and-error basis and it was now time to replan. We needed to replicate natural conditions wherever possible, the most important factor being that the sexes should be kept apart throughout the year until the breeding season arrived. As the adult female is naturally a solitary animal, the 40-hectare enclosure was not large enough to allow them all to go their separate ways, so we decided to allocate to each female a reasonably large piece of land so that she could be alone and have her own territory. In establishing a suitable area along the eastern boundary, Godfrey pegged out eight one-hectare camps. He hired an auger to dig the holes needed for the fencing poles on the stony koppie as it would have taken months to dig the holes manually. This was a great success and when the breeding season came round again in the summer of 1974 we had eight females settled in their own camps. At the same time the males had to be kept out of sight and hearing of the females, so a suitably large area, also within the 40-hectare enclosure but on the other side of the koppie, was selected for them. This became known as the 'Monastery'. It consisted of a tract of level ground at the foot of the koppie where the tall grass was interspersed with bush – ideal cheetah habitat. As we had now learnt that different groups of males are hostile towards one another, three separate enclosures were erected.

Once these new camps in the Monastery had been completed, a day was set aside on which to move the males and Dave agreed to assist us. It poured with rain on the appointed morning but we were determined not to postpone our task. Sopping wet at the end of it all, we were only too happy to see the male cheetahs released and free to run in the spacious new enclosures. An area between the Monastery and the female camps was known as 'No-man's-land'. Here the males would later be allowed to wander at liberty when the females came into season and in this way be free to choose their mates.

That evening while enjoying a well-earned sundowner, Godfrey and I discussed the day's activities and our future plans. During the course of our conversation Godfrey became more serious and said to me, 'You realize, Ann, that we have spent a lot of time and money on erecting all these enclosures and at this stage we really don't know whether our project will work, let alone be successful.'

I thought for a while and replied, 'Yes, that's true, but we won't really know unless we try, will we? Think what a tremendous breakthrough it will be if we succeed. Imagine what a great step forward it will be if, one day, we can breed cheetahs in such large numbers that the species can be taken off the endangered list!'

'I agree,' he said, laughing. 'But aren't you being rather optimistic at this stage?'

He was right and, as I had often done before, I wondered whether our excessive work and expense was to be in vain; but, as always, my optimism outweighed my doubts.

As time went by, Dave became increasingly interested in research aspects and the veterinary approach to the cheetah species as a whole. The field was wide open to researchers as very little – if any – definite scientific findings had been obtained up to that time. To assist him we kept daily records of each animal's condition, and faeces samples were taken from them regularly. Then, one day, through blood analyses Dave discovered that a tremendous amount of fat was not being digested by our cheetahs. On that occasion he showed us samples he had taken approximately half an hour before and we were fascinated to see globules of fat slowly forming on the top of the blood in the test-tube. After further physical examinations his suspicions were confirmed when he found that Anton, a Namibian cheetah that had never been strong and healthy, had an impaired liver – caused possibly by an excessively fatty diet. Dave immediately instructed that all fat was to be removed from the cheetahs' food. This problem, he told us, would never occur in natural conditions, as game – the source of cheetah food in the wild – has very little body fat.

We had observed that the sexual cycles of a number of the females, now in their own camps, seemed to be normal, so Dave felt that it was the males that should be studied more closely. He telephoned us one evening in June 1974 to say that he and Miriam would be visiting us the following morning to measure the males. While he spoke I had visions of measuring in centimetres the length from nose to tip of tail, height from shoulder, width of neck, and so on. When Dave arrived the next day with a pair of callipers instead of a measuring tape, I managed to hide my surprise and asked no questions. Still wondering how he was to cope with this task, we drove the first male into the crush. Then after a few minutes I burst out laughing. After sedating the animal Dave was measuring the size of the cheetah's testicles! He followed this by taking specimens of semen, with an electro-ejaculator, an instrument similar to one used on sheep. Dave was later assisted by Professors Brough Coubrough and Henk Bertschinger, two fellow-veterinarians from the Faculty of Veterinary Science based at Onderstepoort, who gave invaluable service in this exercise to select our best breeding males. Afterwards they took semen samples monthly so that seasonal changes could be detected and any irregularities identified.

The results were surprising. Of the 19 males tested, six were found to have only about 50 per cent of normal sperm. Reasons for this were put down to stress, lack of exercise, and fatty food – factors we now set about eliminating. We felt the problem of stress would automatically improve with the animals being in larger enclosures. To combat lack of exercise, a 'racetrack' – an area cleared of bush where the cheetahs could run unhindered – was planned. Apart from removing the fat from the chickens and red meat which we fed them – a process we had already started – the zoo agreed to buy rabbits as food as they were high in protein and had little or no fat on their meat. Interestingly, Dave subsequently tested cheetahs in the wild, and their semen showed similar results.

Only the six cheetahs that had managed to pass the semen test were to be used for breeding but the others remained with them in the group. Among the unsuccessful males there was old Danie, estimated to be a 12-year-old, who was scratched and scarred, not particularly interested in sex, but determined to remain the Boss. We realized that he was of no use to our breeding project but we had decided that he would spend the last few years of his life in peace on the farm.

At his age he would not last long if returned to the wild, so Danie remained in charge and was exempted from further semen tests.

Of all the Namibian cheetahs, Fanus was my favourite. He was not tame, but not aggressive either. Even as a young animal, he went his own way, his only companion being Tom. Fanus's beautiful brown eyes were always alert, his coat shone, and his muscles were taut: he was a fine specimen. But according to the semen tests, he was no good for breeding. Chris, on the other hand, was a large animal, very scruffy and battle-torn. If looks were to influence our selection for breeding, Chris would have been near the bottom of the list. He turned out to be our finest fertile male and later fathered a good number of cubs.

Our poultry business was now expanding rapidly. The chainstores that we supplied in Pretoria were opening branches in new suburbs and demanding an ever-increasing quantity of eggs. This needed full-time supervision and as the cheetahs were taking up more and more of our time, Godfrey and I decided to employ extra staff. A farm manager, Dennis Crowther, was appointed to help with the running of the farm and someone was hired to organize the egg-packing station. From the start Dennis – or 'Denny' as he was known – took an interest in the cheetah project and helped wherever he could. I was pleased with this extra assistance as I could now spend even more time observing our cat family and learning about them. I was required to help out in the egg-room only during rush periods.

Early in 1975 we noticed that some of the females had become secretive: they did not come as usual to the gates of their camps at feeding time and we felt sure this was a sign that they were coming into season. We released one group of males into No-man's-land and in the afternoon of the same day we discovered on our rounds that they had already found their way to the females. They wandered up and down the worn track that passed all the female camps and as they did so we could easily observe if interest was shown on either side. The bullies of the group continued to charge the females at the fence, but fortunately the selected breeders were not very aggressive. I decided to watch their movements and interactions in the early mornings and late afternoons as in the heat of the day the cheetahs were lazy and very little activity took place. I learnt a great deal about these secretive cats during this period and realized that they were all very different to each other in character and mannerism. With some of the females it was obvious that they were in season, but with others it was almost impossible to recognize the signs. Jill, for instance, did not roll on the ground or come up to the fence to greet the males. It was only because she watched the males intently from a distance that I decided to release Frik into her camp. He was very lethargic, taking some time to 'read the newspaper', and by the time that he eventually met up with Jill it was dark. I left him in her enclosure, knowing that as the two of them had grown up together, he would not attack her. After a few days he was removed from her camp, but at no time was mating observed.

Lady was just the opposite – she was definitely in season. I watched her come up to the fence and sniff at the males, nose to nose, through the wire mesh. Then she would roll over onto her back, right in front of them. I released Al into her camp but there was no attentiveness shown on his part. Al was removed and Chris introduced, whereupon he immediately mounted her.

Later I had the same experience with Kromstert. She was not at all interested in Al who was wandering about her quarters, but ran to greet Chris the minute he entered her enclosure. After a certain amount of chasing, she allowed him to mount her. Chris definitely had something, we concluded: he was certainly a great favourite of the ladies! On another occasion Purry was at the peak of her oestrous cycle; she performed in front of the males and I let in one of the selected Namibian group. Without warning, the male suddenly attacked and bit her on the hindquarters and it took some time to chase him out of her enclosure. In the meantime Purry had disappeared into the bushy area of her camp and had lost all interest in the opposite sex. Fortunately her lacerations were not serious and 10 days later she again showed signs of coming into season. This time I took no chances and released Chris into the enclosure with her. After the usual sniffing ritual, Chris located Purry and mating duly took place. I realized how experimental our work was and how much we still had to learn.

Snippie the caracal was pregnant again. The gestation period was only 72 days and she was close to the end of her time. Having separated her from Oubaas, we left her warm and comfortable in a small, specially built, grass hut. Snippie greeted me at the gate early one morning and to my intense delight there was no repetition of the earlier episode for she led me to her new-born brood. She had given birth to three kittens, two of which were healthy, but the third still-born. In the past I had hand-raised many domestic kittens from birth, but by comparison there had never been any as handsome as these. They were bundles of dark golden coloured fur with long tufts of hair on their pitch-black ears. Eyes closed, crying plaintively, they crawled on their bellies. Although Snippie accepted my presence, she still seemed very much on edge and the kittens, too, were restless. I sought

Dave's advice and he told me to stay with her, watching to see whether the kittens were drinking: it was possible that Snippie had no milk. It did not take long for me to discover that this was indeed the case, and so I took Snippie and her two kittens down to the house where the youngsters could be fed artificially. In the meantime, as Snippie's milk-line was good, Dave gave her an injection to help bring on the milk. The kittens hungrily accepted the bottles that we offered them; Dave had obtained small bottles with tiny teats, used normally for premature human babies, and had worked out the formula for the milk mixture. Little did we know that this experience was perfect practice for what was to come.

March 1975 was a tense month for us all. We were aware from the size of the abdomens of the cheetah females that six were pregnant, but could predict the exact birth-giving dates of only three – those of Lady, Kromstert and Purry. Naturally we feared a repetition of the previous year's disappointment of false pregnancies. And if cubs were born, we wondered whether the mothers would look after their young for we had heard that if they were under any stress they were inclined to eat them. At the same time, there were several wildlife experts who were very sceptical and pessimistic about our chances of success. However, Frank had said to me earlier, 'Ann, my zoo council and I are behind you all the way. After all the time, expense and effort you and Godfrey have put into the project, we know that it cannot fail.'

This encouragement was all we needed.

Visits to Pretoria had become very few and far between, but early in April I had to attend to a business matter in the city. I asked Denny to cope with the feeding of the animals as Godfrey was busy in the egg-room. On my return to the farm late that afternoon, Godfrey and Denny greeted me at the steps that led to the front door of our home.

'Ann, we have wonderful news for you,' said Godfrey, with a smile. 'Jill has given birth to cubs.'

I was completely taken aback and faltered, 'You're joking!'

'No, Ann,' Denny replied. 'Although I've not actually seen them I definitely heard them crying when Jill came for her food.'

I simply could not believe it – what we all had battled unsuccessfully to do for the past four years, had finally been achieved. Had we made the breakthrough? Would there be other births? Were we at last on the road to success? My immediate thought was that Dave and Miriam should be told the wonderful news.

(14) *Although not adept at climbing, cheetahs often take up a position on the lower branches of trees.*

(15) Mof displays a typical stance of aggression.

(16) Group cohesion is strengthened by face licking.

(17) Tana.

(18) Lady.

(20) *Caracals Oubaas and Snippie warming themselves in late-afternoon winter sunshine.*

PREVIOUS SPREAD:
(19) *Mof and Al on arrival from Namibia.*

(21) *Oubaas, indigenous to the Magaliesberg range, had been found on our farm.*

(22) Snippie with her ten-week-old kitten.

*(23) Unlike cheetahs, caracals
have no tear-lines
and their ears are tufted.*

(24) Frik sitting in the tall grass shows how well cheetahs blend in with their natural surroundings.

'Yes,' Godfrey said, as if reading my mind. 'We thought of them, but wanted to tell you first. Hannes and Frank won't be in their offices at the zoo now, so we'll telephone them early tomorrow morning.'

Needless to say Dave and Miriam were as excited and as thrilled as we were and although we could not be together, champagne corks were later popping in both homes.

Driving back from Pretoria late that afternoon I had noticed storm clouds building up in the west. There had been a cold nip in the air and I had surmised that if rain fell it would probably herald the winter. Before I retired to bed a light drizzle had started to fall and to make matters worse the wind had turned exceptionally cold. Winter was setting in. I slept very little for my mind kept wandering to Jill's camp. Was she protecting her cubs from the wind and rain? Just our luck, I thought – with the very first litter. Then I imagined the cheetahs in the wild where this sort of thing must surely also happen, and I began to feel more at ease. We had all agreed not to interfere with the mother and her cubs and, unless absolutely necessary, never to remove the young from her care. We needed to breed self-sufficient animals for the wild, not dependent pets.

Early the following morning, after the rain had subsided, Denny and I took Jill's morning meal to her, hoping for a glimpse of the cubs. But we were most perturbed by what we saw: lying on either side of the separating fence, close together, were Jill and Helga, Jill being some distance from where Denny had seen her the evening before. There was no sign of any cubs and – if they were still alive – she showed no interest in them. My heart sank. Had we battled in vain? Had we found a way to breed cubs only to find that the mothers lacked maternal instincts?

Hurriedly I followed Denny into the camp and he led me to the area where he thought the cubs should be. We searched for a few minutes, all the while noticing that our presence drew no reaction from Jill. Then I heard Denny say, 'Oh, Ann! I'm sorry.'

I immediately went to his side and there, lying scattered in an open patch of the camp in front of him, were three lifeless forms. I could say nothing. Kneeling down I picked up one of the motionless cubs to look at its face, something I had waited so long to see. Its body was cold and stiff and the fur wet through. Its little mouth was blue.

'Why, Denny, why?' I exclaimed vehemently. 'They can't be dead!' I cried, refusing to accept what seemed the obvious. Then as I held the cold little body against my jersey, I felt a slight movement.

'They're alive!' I yelled at Denny, and grabbing all three, I pushed them under my thick pullover and rushed back to the truck.

CHAPTER 6

The Breakthrough

*T*he bakkie bumped down the uneven road at top speed. I was conscious of the three cold and wet bodies against mine and as I looked down I heard a slight whimper from one. Godfrey, out on the lands, saw and heard the rapidly approaching vehicle and immediately dashed to the house to assist with what was obviously a crisis. Soon two electric heaters had been brought out and, seated close to them, I tried to dry three bundles of very wet fur. My rubbing increased the circulation and with the warmth from the heaters, gradually the tiny animals came back to life. It was incredible how the cold, lifeless lumps of flesh miraculously became warm, soft and hungry youngsters. Having never seen cheetah cubs at such close quarters before, I now had the opportunity to study them.

At this age the only marked resemblance they have to their parents is in the tear-lines on their faces. The mantle which covers the back of the cub is short at birth but grows quickly into thick, soft grey fur which extends from the head to the tip of the tail, while the lower parts of the body are very dark – almost black. If one parts the fur, cheetah spots can be seen repeated on the animal's skin. Reasons have been suggested for this unusual coat – one being that it resembles that of a honey badger or ratel, a ferocious mammal that has no fear of larger predators and for which the latter have great respect. The mantle disappears by the time the cub reaches the age of six months, leaving just a slight mane between the shoulder blades at the base of the neck and the spots become prominent. If the fur is part of its camouflage, so too is the cheetah cub's high-pitched bird-like chirrup. This call can easily be mistaken for that of a small bird – a mossie, or sparrow. These tiny wriggling, furry animals are extremely vulnerable, not only to predators but also to disease, and we all realized that our challenge now was to keep them alive. Dave had of course been told of the unfortunate turn of events and he was soon on his way to help: in no time at all he and Miriam were on the farm. While we were all saddened that Jill had abandoned her cubs, we were nevertheless thrilled that we had the three youngsters alive, albeit in human care. Cautious as always, Dave warned us that the next three weeks would be the most critical. He formulated a mixture of fresh cow's milk and egg yolk which the cubs, now warm and very hungry, gulped down greedily, purring continuously as they did so. Meanwhile in the room in our home that by now had been set aside specially for such emergencies, Godfrey set up the infra-red lamp that Dave had brought with him. Dave warned us that, as orphans, the cubs could not regulate their own body temperature for the first few weeks of life, and a constant temperature of 31 to 32 degrees centigrade would have to be maintained artificially, especially during the approaching cold winter nights.

After the young felids had drunk their fill of the formula, Dave lifted them one at a time and very gently, with a warm, damp wad of cotton wool lightly rubbed each one's tummy. In no time each cub had urinated, and with a little more pressure, had defecated too. This we learnt was something that would have to be done regularly. The mother would normally have activated these processes by licking her offspring's bellies with a firm, wet tongue. The cubs were then weighed and sexed: there were two males, each weighing 500 grams, while the only female weighed in at a grand 552 grams.

'Ah, perfect!' noted Godfrey, on hearing the combination of the sexes of the cubs.

'We'll call them after Dave and Miriam's children, two boys and a girl – Erich, Mel and Cornell.' And so they were named.

One of the males, Mel, had been badly bruised at birth and, in order to counteract infection, antibiotics were prescribed for him. Frank Brand had told us that cheetah cubs are very susceptible to pneumonia, so the other two received antibiotic injections as a precautionary measure. With tummies now full and the stress and strain of their ordeal over, the orphans soon fell fast asleep close together, a single mass of soft grey fur. In spite of the obvious difficulties that loomed ahead, it was with a wonderful feeling of achievement and fulfilment that we watched the three helpless youngsters, their Roman noses twitching in their sleep, and knew that from now on they were totally dependent on us for survival.

That afternoon, Jill was caught in a crush: when sedated, Dave took milk from her which the cubs were to drink with great enthusiasm later. He also took blood samples which were tapped into test-tubes. These were placed in a centrifuge and spun at a very high speed to separate the plasma from the red cells. Five millilitres of this plasma were injected into each of the three youngsters. Dave explained that in all mammals, during the first 24 hours after birth, the female has in her milk a thick creamy substance known as 'colostrum'. This contains a high concentration of antibiotics and Vitamin A which protect the offspring from infection and disease during the early stages of life. In the case of Jill, however, it was now too late to obtain this milk, so as a substitute the plasma from the blood which Dave had injected into the cubs would have the same effect as the colostrum. He insisted that we should keep records of the milk intake and daily weight-gain of each cub and instructed us to feed them every three hours throughout the day and night. After being fed, their bellies were to be massaged, stimulating them to urinate and defecate.

I have never been a light sleeper, so I found the early morning three o'clock feed very difficult. When the alarm clock sounded, I had to force myself out of a cosy warm bed and then stagger sleepily to the kitchen to warm the cubs' food mixture. Laden with thermos flasks, freshly washed bottles, cotton wool and basins, I would stumble into the cubs' room where the sleepiness quickly dissipated when I was rewarded with the loud purring of three hungry youngsters. Having measured out 20 millilitres of their food mixture in the small baby's bottle – they usually drank between 10 and 20 millilitres each – I started by feeding Cornell. When she spat out the teat, indicating that she had had enough, Erich and Mel would have their turns. On each of three charts I recorded the time of feeding, the amount taken and whether or not the cubs had urinated and defecated.

As each cub finished feeding, it was handed to my dog, Girlie, who took over the duty of performing their toilet. As the mother cheetah would have done, she licked and groomed them, seemingly with great motherly love and ostensibly accepting them as her own offspring. Her loving nature and strong maternal instincts were to be of great help to us now and in the future.

Girlie had joined us about a year before this. Obviously abandoned, this stray young bitch had been picked up on a highway outside Pretoria by a friend who

Table 1 CHEETAH CUB FEEDING CHART

PARENTS: JILL AND F.R.I.K. . CUB IDENTIFICATION: . CORNELL . . . . SEX: FEMALE . . . . . BIRTH DATE: 4/4/1975

DATE	TIME	MASS	INTAKE	URINE	FAECES	REMARKS
4/4/75	7.15					ABANDONED BY MOTHER TAKEN AWAY COLD AND WET RUBBED BRISKLY TO DRY AND INCREASE CIRCULATION
	7.45	552g	20ml	✓	✓	FED MIXTURE OF COW'S MILK AND EGG YOLK ANTIBIOTIC INJ. AGAINST PNEUMONIA
	10.45		10ml	✓		
	13.45		10ml	✓		5ml BLOOD PLASMA INJECTION
	16.45		10ml	✓		JILL'S MILK GIVEN ORALLY
	20.00		7ml	✓		
	23.00		14ml	✓	✓	
5/4/75	2.00		13ml	✓	✓	
	5.00		12ml	✓		96ml TOTAL INTAKE FOR 24 HOURS
	8.00	576g	15ml	✓		
	11.00		10ml	✓		
	14.00		15ml	✓	✓	
	17.00		20ml	✓		
	20.00		16ml	✓		
	23.00		12ml	✓		

DATE	TIME	MASS	INTAKE	URINE	FAECES	REMARKS
6/4/75	2.00		14ml	✓		
	5.00		15ml	✓		117ml TOTAL INTAKE FOR 24 HOURS
	8.00	600g	7ml	✓		
	11.00		8ml	✓		
	14.00		15ml	✓		
	17.00		16ml	✓		
	20.00		19ml	✓	✓	
	23.00		11ml	✓		
7/4/75	2.00		16ml	✓		
	5.00		18ml	✓		111ml TOTAL INTAKE FOR 24 HOURS
	8.00	620g	20ml	✓		
	11.00		20ml	✓		
	14.00		21ml	✓		
	17.00		20ml	✓		
	20.00		23ml	✓		
	23.00		23ml	✓	✓	
8/4/75	2.00		18ml	✓		
	5.00		22ml	✓		167ml TOTAL INTAKE FOR 24 HOURS

asked me to adopt her. One of Heinz's 57 varieties, Girlie was the size of a large fox terrier but her features, tail and colouring were similar to – and her intelligence certainly matched – that of a sheepdog. Judging by her teeth, we thought she had probably been not more than three years old at the time. Girlie and I had taken to each other immediately and Godfrey had been quite happy to accept her into our home, but on one condition – that she be spayed. I had smiled to myself at the time. 'Dear Godfrey,' I had thought. 'He really will take a tent and pitch it on the other side of the koppie if Girlie ever presents us with a litter of mongrel pups.'

And so Girlie became a loyal friend and followed me constantly, either on foot or in the truck, all over the farm. She had no fear of the cheetahs and seemed to feel that when I was with her there was no danger. If I brought an abandoned animal into the house to be hand-raised, Girlie seemed instinctively to know how to care for it.

Our attention turned to the other pregnant ladies. Jill's pregnancy had, in fact, been a surprise for it had not been obvious to me when she had come into season and I had not been at all sure if she had been covered. Now we wondered whether there would be other mothers that would abandon their young. With our time fully occupied attending to and feeding Jill's litter three-hourly, we hoped that we would not have to care for further abandoned cubs. How complacently we had sat back after the mating season was over, thinking that with that accomplished our biggest problem was behind us and little realizing that young cheetah females could be such poor mothers. Our other mother, Snippie, had by now been taken back to her enclosure as she was lactating well and had sufficient milk to feed her kittens. Her mate, Oubaas, had been placed in an adjoining camp as we feared he would kill the youngsters.

The Pretoria Zoo council and personnel were thrilled with our success. How well I remember Frank's telephone call early on the day following the birth of Jill's three.

'Ann, powder your nose – I'm bringing out the Press this afternoon,' he informed me.

Laughing, I agreed to do so, expecting that he would have no more than two or three reporters with him. I had just finished feeding the cubs when Godfrey called me to join our visitors. Walking into the sitting-room with a large cardboard egg-box containing a baby's blanket that Miriam had given me and three small furry objects cuddled up into a single ball in the centre, I was stunned to find the room crowded to its limit. Press photographers, feature writers, TV cameramen –

you name it, they were all there. It was good, however, to see the familiar faces of Frank, Dave and Miriam among the sea of strangers.

'How do you account for your success, Miss van Dyk?' 'What's the secret?' 'Do you expect any more?' 'Did you deliberately take them away from the mother?' 'Why were they abandoned?' 'What made you start on this venture?' I was experiencing my first introduction to the media and I was nervous and tongue-tied but fortunately Frank came to my aid and answered most of the barrage of questions. Godfrey and I sighed with relief when the last visitor had left, and with Dave and Miriam as well as Denny – who had completed the feeding of the adult cheetahs in the meantime – and Vi, Denny's wife, we all sat down on the veranda to enjoy a sundowner and to chat about the exciting events of the day. Once the news was out, the telephone never stopped ringing for there were calls of congratulations from friends and well-wishers local and far afield. All the while the cubs slept soundly and Jill, seemingly unconcerned, roamed her territory, unaware of her offspring's importance as one of the few known litters of cheetah cubs to have been bred in captivity.

Our family of three developed quickly. They continued to gulp down their food and became rotund, cuddly and playful. They learnt to react to human voices and tottered and stumbled across to me on shaky legs when I called 'Kossies!' at feeding time. While kneading with their front paws against my hand, they would suckle from the bottle. I felt their claws, sharp and pointed like a cat's, but knew that these claws would wear down and become blunt as they grew older.

A new problem now arose among the adult cheetahs: it started with Frik and Mannetjies. We noticed that around their eyes they had small dark hairless patches and over the next few days, these areas enlarged in size. On examination and after taking smears, Dave diagnosed mange, an infectious skin disease caused by external parasites that possibly had been introduced to the farm by the rabbits on which the cheetah had been fed. The disease spread rapidly and all animals that showed the symptoms were immediately sprayed with an insecticide that was known to control the mites. Meanwhile, a close watch was kept on the healthy cats. Treatment was repeated two weeks later and gradually hair started to grow on the infected areas. Fortunately it was not necessary to spray the five expectant mothers for they showed no sign of the disease.

We were now experiencing what seemed like the calm before the storm: we did not know quite what to expect from our five possibly pregnant females. We wondered how many pregnancies would be false and how many mothers would care for their young. The abdomen of each one was swollen and hung low, well below the rib cage. Mammary glands were enlarged, forming a milk-line, and teats could be seen through the fur. I fed the expectant mothers larger portions of meat daily and gave them a bowl of milk every morning. In the wild, during and towards the end of her pregnancy, a cheetah female has to build up her physical condition and stamina to cope with the period prior to giving birth. Towards the end of the gestation period, her heavy and swollen body is a great hindrance and retards her speed when she runs down prey, often leading to unsuccessful attempts to kill. When the time for giving birth is due, she will select a secluded and protected lair in tall, thick grass or in dense, thorny bush where she will be concealed and reasonably safe from other predators. Litters vary in number from three to eight cubs at most. After the birth of her young, using her teeth the mother will gently break the membrane enclosing each offspring and then lick the young animal clean. This membrane, together with the placenta, or afterbirth, is eaten by the mother, thus leaving no trace of blood or smell to attract meat-eaters. The consumption of the placenta also provides extra nourishment to the female who is physically unable to hunt so soon after giving birth. Lying on her side and always alert, she moves very little and allows the blind and defenceless new-born cubs to find her teats by themselves and feed from her. To prevent other carnivores from scenting out her offspring, the mother will move her brood continually during the first few weeks after birth. One at a time, she carries each cub gently in her mouth by the scruff of its neck and takes it to a new and clean lair.

Dawn was the first of the five to refuse food – usually a definite sign that she would soon be going into labour. On visiting her one morning – which happened to be a Sunday – I could hear the plaintive call of a cub near the entrance to the encampment. Peering through the fence, I saw Dawn relaxed and lying in the grass and about two metres away from her lay a crying cub. I sat still and watched her for some 10 minutes during which she made no attempt to comfort or feed her offspring. Once again I had no alternative but to remove the cub from its mother. With Godfrey and Denny keeping Dawn at bay, I lifted the wet, cold and miserable creature from the ground. At the same time, I noticed that not far off lay a strange object. No, surely not! But yes, it was! Only the head remained of another cub – Dawn had eaten the rest of what had probably been a still-born cheetah.

Quickly we drove to the house where this time we were better prepared for the emergency. The heaters were on, the room was warm and the milk mixture was already prepared and in the fridge. The new cub, a male, and later called 'Stephen', was in very poor shape and weighed only 360 grams. He had no hair on his body and his skin was hard and leathery. But like the others, after a little coaxing he took to the bottle. Dave was soon on hand to give Stephen an antibiotic injection as he had done with the others.

As Dawn still looked pregnant, Dave examined her in a crush and ascertained that there was still one cub to come. Three days later Dawn went into labour again and gave birth to another cub which was physically quite normal. As such a long period had elapsed between the births, it was thought possible that the first two had been premature. But once more Dawn simply watched her latest offspring from a distance and made no attempt to feed or keep it warm. Oh dear! Here was yet another cub to join our ever-increasing litter at home. It was ironic that we had wanted cubs so badly and now we were certainly getting them, but in the wrong circumstances. I was feeding Stephen three-hourly and the new arrival, whom we called 'Dettie', joined the routine. Girlie took all this in her stride and again appointed herself as mother of the brood. Dettie was a very thirsty cub and drank her milk mixture quickly. Then, with a full stomach, she sleepily joined her brother Stephen in the box that we kept them in and they snuggled together for the night.

A few days later at feeding time Helga was missing. We now realized that probably because they were young and these were their first litters, the inexperienced females had abandoned their cubs readily. With this in mind we decided to look for Helga. While searching her camp, Denny stumbled on her unexpectedly. Hidden in the long grass, crouching low, she lunged forward when he approached, raising her front paws and bringing them down again with lightning speed, fortunately just short of his body. With adrenalin surging through his system, Denny made a hasty retreat, reaching the gate just in time to escape Helga. For the rest of the morning Joel was stationed in a tall tree outside the camp to observe whether any cubs appeared and if so, if they were suckling. He later confirmed that there were four youngsters, all strong, healthy and drinking from Helga. At last we appeared to have one good mother.

Erich, Cornell and Mel were by now six weeks old. Approximately 10 days after birth their eyes had opened; at three weeks their milk teeth had started pushing through, reaching a full complement by the age of six weeks. Now off the bottle, they were lapping their milk-and-egg mixture and it was time to wean and start giving them solid food. On instructions from Dave, I gave them a very small portion of finely minced raw beef once a day. It was amazing to see how they soon recognized the smell of the meat and devoured it greedily. A week later I added a little softened commercial puppy chunks to the mince, as well as extra calcium and vitamins. After taking one sniff at this concoction, the youngsters turned away and looked at me quite startled.

'You expect us to eat that?' they seemed to say.

So I skipped one meal and when feeding time came round again they gobbled up the new combination hungrily and without any hesitation. I slowly increased the quantity of artificial food and mince until they were eating a well-balanced porridge-like mixture three times a day.

Captive cheetah cubs require extra calcium, copper and thiamine in their diet. The calcium is to counteract rickets, a bone disease; the copper prevents sway-back, a weakness in the hindquarters; and thiamine, or Vitamin B, suppresses a nervous disorder. Under natural conditions, when her cubs are approximately six weeks of age, their mother selects a young animal as prey and after killing it calls and leads her offspring to the carcass. Tearing it open, all the while uttering low guttural sounds, she encourages her young to nibble off small pieces of meat and to lap up the fresh blood. In this way needed vitamins are obtained from the kill's soft ribs, cartilage and internal organs such as the liver, kidneys and heart; seldom are the stomach and intestines eaten.

The cubs soon become very active and playful, stalking, chasing and tripping one another in what appears to be a rough-and-tumble game, slowly but surely in so doing, learning the killing techniques. But it is only at the age of approximately six months that the mother will encourage her cubs to accompany her on the hunt. Until that age they are a hindrance to her for, being very boisterous, they may give warning of their mother's presence to the selected victim. At this early stage, however, the mother sometimes catches and brings back small live prey for her brood to kill. This, together with the 'games' they play, conditions them for survival in the wild.

By the middle of May, winter had truly set in. Over the previous few weeks the countryside had slowly and almost unnoticeably changed and the warm, subdued autumn-coloured leaves had fallen, leaving most trees stark and bare. A dry wind blew softly over the corn-coloured grass, dropping seed onto the ground for ever-busy termites which were packing their already overstocked larders for the cold months that lay ahead. Lady and Kromstert still had to give birth and we wondered whether they would attend to their young and keep them warm during the chilly nights. In anticipation, we erected pyramid-shaped grass huts for the two of them – hoping that the warmth that the dens provided would entice the mothers to give birth inside them and so shelter and protect any new-born cubs.

Having left Helga in isolation for the past few days, Dave felt that he should check up on her cubs. Judging by the many patches of flattened grass within her camp, she had obviously moved her young a number of times. She was so well camouflaged in the bush that, like Denny, Dave came upon her unexpectedly and Helga reacted in the same manner, jumping at him, catching him with her sharp dew claws and tearing at his clothes. Her cubs were not far off, but there were only two. Where were the others?

Giving Helga a wide berth, we found one dead cub in a clearing some distance from her. There were small tooth-marks on its body – probably those of a genet, a small indigenous cat that on rare occasions Godfrey and I had spotted at night while driving along our farm roads. Of slender body and with a long tail, it glides silently through the bush and long grass seeking refuge in the nearest tree when danger lurks. It has a long pointed nose and rounded ears, a light silvery grey coat covered with black spots and blotches, and rings of dark fur round its handsome tail. When the genet is annoyed or cornered, a black stripe of long hair down its back is raised and, snarling and hissing, it will then lash out. It is a nocturnal animal and its varied diet consists of small mammals, birds, eggs, reptiles and insects. We surmised that it was a genet too that had eaten the other cub and so we decided to keep a closer watch on Helga and the rest of her litter – but from a distance.

For the next few days all went well. Then, as if Helga had suddenly become bored with the idea of being a mother, she lost all interest in her brood and one day we found her lying some distance from her crying offspring. Nightfall was approaching and once again we had no option but to remove the remaining two. We managed to revive one of them, a female later known as 'Di', but her brother died that night of pneumonia. With six cubs to feed now, everyone at De Wildt Estate helped whenever and wherever possible – including Girlie who busily washed cubs' faces after every meal and generally helped with their toilet.

Exactly 93 days after Lady had been covered by Chris, she refused to eat and retreated to her grass hut. She remained inside for the next couple of days – a sure sign that she had given birth. Twice a day we placed her food next to her water-bowl and then left promptly as she had to be kept as isolated as possible. A week later she came out for her meat when we approached and to our delight we could catch glimpses of four healthy youngsters sleeping on the floor of the triangular-shaped grass hut. We felt confident that Lady would not abandon her cubs for she was extremely aggressive and clearly protective of her brood.

Kromstert, our remaining pregnant female, was very big-bellied and looked decidedly uncomfortable. She refused to eat one afternoon and we expected cubs the next day. While taking Lady's meat and milk to her early the following morning, Denny and I stopped on the way at Kromstert's camp. We heard a cub crying – but there was no sign of the mother. It had been bitterly cold the night before so without hesitation we set out to look for the cub. While doing so I came across a large flattened area of grass and stared at the objects there in disbelief. Scattered on the ground lay four youngsters, still in their birth sacs – Kromstert had made absolutely no effort to complete the birth process and to clean her new-born cubs.

Without saying a word to each other, Denny and I quickly bent down to remove the thin membrane covering the cubs – but we were too late. They had all suffocated. One was small and obviously still-born, but the other three had been beautiful, normal cubs. This, I felt, was the last straw! But we had no time to brood on the misfortune for we suddenly heard the weak cry of what was to be the only survivor of the litter. It was a very faint call and it took us a few minutes to locate the cub. When we finally tracked it down, we found that this youngster, too, had not been cleaned, but at least the sac had been broken so that it could breathe. Yet again, we hurried to the house where Denny rubbed down and cleaned the cub with warm water and cut the cord while I sterilized a new bottle and warmed the milk.

The number of abandoned cubs still totalled six as Dettie had died a few days previously from heart malfunction. Although the results of these first births had been disheartening, we had learnt an enormous amount and we knew that a number of changes would have to be made before our next litters arrived. That afternoon Dave came to take blood from Kromstert; the sample was spun and the separated plasma then given to the new cub together with the usual antibiotic injection. Later, having fed our family, Dave, Miriam, Godfrey and I sat down with our sundowner on the veranda. Suddenly there was a scuffle in the kitchen and on investigating we found that the farm dogs had chased a beautiful, tame, but sickly looking domestic cat behind the stove. Dave was very worried as it obviously had some infection and our cubs in the adjoining room were dangerously close. He decided to take the hapless cat back to his Pretoria surgery where he could conduct tests to determine what was wrong, and treat it accordingly. However the cat died two days later and, looking back now, it seems that this was a forewarning of the problems that stray domestic cats would bring. Fortunately in this instance no infection was passed on from the sick animal to the young cheetahs.

The weeks passed rapidly and just as quickly the cubs grew in size. Soon a large area in the garden had to be camped off for them, with a passage from it that led into one of the rooms of the house. The nights were still cold so the cubs slept indoors. They loved this new arrangement; now they could stretch their long legs when they chased one another and they enjoyed jumping onto the vantage point of the lower branches of a marula tree. Cheetahs have excellent eyesight and hearing, but their sense of smell is poor and for this reason they take every advantage of the height of a koppie or anthill when selecting their likely prey.

Lady remained the one and only perfect mother that year. Attentive and caring, she was very protective of her family. When they reached approximately four weeks of age, she began to encourage her young to emerge from the den. While she lay outside in the sun, she coaxed them to join her by uttering soft bird-like calls. Timid at first, they crept out cautiously on their bellies, but at the slightest noise scurried back to the safety of the hut. Gradually they became more adventurous and explored further and further. They clambered over Lady's body, jumped on her flicking tail and purred while she groomed them gently with her rough tongue. On the surface it seemed to be a quiet and peaceful scene, but Lady was always tense in her alertness, gazing far into the distance at times, as if searching for possible predators. At feeding time she began to call her cubs to share her portion of meat with her. I now had to devise a way of preventing the youngsters from eating the meat as they had to be weaned onto the same well-balanced mixture that I was feeding to our hand-raised group. I decided to cut Lady's chunk of beef into small portions making it impossible for her to carry the meat to her offspring, and while she gulped down this food, I kept her cubs at bay. When she was finished I placed in the camp a bowl of the recommended cub mixture, which the youngsters, with encouragement from Lady, gradually learnt to lap up.

The Leopard Story

One Wednesday, a day that Dave had set aside every week for work on our farm, we were spraying a few cheetahs that still showed signs of mange. After catching the animals, one at a time in a crush, we doused the cats thoroughly with the aid of a bucket and stirrup pump. They spat, twisted and turned in the confined space and showed themselves to be thoroughly opposed to the soaking. As with other members of the cat family they hated being wet. The normally stoical felids looked most undignified and bedraggled sitting in the crush with water dripping from their heads and bodies. But as the rest of our cheetah family were now clear of the disease we wanted to stamp it out completely – and so their feelings had to be ignored. While we were busy with this chore, a woman from a neighbouring farm suddenly appeared and, gesticulating excitedly, she babbled forth in her home tongue.

'Aw, aw,' was the answer she got from Shironga. Not understanding what the woman said, Godfrey turned to Shironga and asked, 'Shironga, what's wrong? Why is she so upset?'

He replied: 'She says that while out collecting wood this morning, half-way up the mountain she saw a cheetah.'

'That can't be, Shironga,' I exclaimed. 'You know that when we checked earlier on, all the cheetahs were in their camps.'

79

'All the same,' Godfrey said, 'we'd better make doubly sure. Ann, you go with Shironga and recheck, while Dave and I carry on here. Shironga, thank the woman for telling us and ask her please to wait.'

Within half an hour we were back, reporting that all the cheetahs were accounted for.

'Godfrey,' Dave said, 'this woman is genuinely upset. Let's quickly finish the spraying and go with her to where she saw the animal.'

Later as we climbed up the kloof on an adjoining farm, we met a farm labourer accompanied by his dog, both descending hurriedly and the man mumbling to himself. He breathlessly told us that a big cat had lunged at his dog, but on seeing him it had disappeared into the bush.

'It must surely be a leopard,' Godfrey said. 'They are known to be partial to dog's meat and only leopards can live wild on this mountain. This isn't cheetah habitat.'

After comforting the obviously scared man, we asked him to lead us to the spot where he had encountered the large cat – at the same time showing him the dart gun that Dave carried to sedate any animal that we might find as well as the shotgun that Godfrey had with him for emergencies. Eventually the man agreed, albeit reluctantly.

It was quite a steep climb. The bush was thick and the old moepel trees formed a dense arch of foliage over the kloof. This in itself was frightening, and I kept looking up into the overhanging branches. Suddenly there was movement in the grass in front of me and I jumped back, only to laugh at myself as a hare went scampering by. Finally we reached the spot where the man had last seen the animal. There we observed leopard spoor in the sand, and for some time we searched for the cat, even though we presumed that by then it had gone much further up the mountain. After calling off the search we visited the owner of the land and asked him not to shoot the leopard if it was sighted again, but to contact us instead. To this our neighbour agreed readily. But the leopard is a very secretive cat and like all wild animals avoids contact with humans. This one appeared to be no exception. Its encounters with people must have made it wary of venturing down into the valley and as far as we know it was not seen in our area again.

That evening when I was feeding my family, I noticed that Kromstert's cub appeared to be weak and wobbly on its legs. She had eaten very little so I immediately sought Dave's advice. He told me that he could not leave his practice at that moment, and that I should keep the cub warm until he was able to come out and look at it early the following morning. However, her condition worsened and shortly before midnight Godfrey and I decided instead to take her to Dave's surgery in Pretoria. On examination Dave decided that she had a blockage in the intestine and gave her an enema. Nothing was forthcoming, so she was given liquid paraffin by mouth. She was put on a drip and we returned home at three in the morning, leaving the cub sleeping next to Dave's bed for the rest of the night. Soon after we left, Dave told us later, her bowels worked and by daybreak she was already showing signs of recovery and was able to return home. We had removed a kitten from Snippie's recent litter to keep this cheetah cub company as all the others in our care were much older and bigger. It was touching to see the welcome the cub got from the little caracal when Dave brought her back to the farm. They licked each other's faces and scampered around the camp together. But the cheetah cub was still weak from its ordeal and the two of them soon curled up together, a single spotted golden ball of fur.

Sue Hart had visited us a number of times since she had treated Tana and in conversation had mentioned a new game-farm, Londolozi, situated in the Sabie Sand Game Reserve bordering the Kruger National Park. One day she invited Godfrey and me to accompany her there the following weekend. As we could not both leave the farm, Godfrey suggested that Denny should take his place and go with me. The morning that we left, however, I was a little concerned about abandoning my brother with no assistance as the caracal kitten had vomited during the night. Godfrey told me not to worry and assured me that the kitten's illnesss was probably due to overeating. Dave was away at the time, and I certainly left with misgivings. However, on telephoning Godfrey the following day I was told that the kitten had improved and was eating a little, and so I relaxed and turned my mind to enjoying the break.

Sue, Denny and I spent a tremendously exciting few days with the Varty brothers, owners of Londolozi which is a private game reserve. We were accommodated in an attractive and cosy thatched chalet, and game-spotting drives were arranged during the day. In the evenings we joined the other guests for a braai at a large open crackling woodfire. On one of the daytime drives in an open Land Rover, we came upon a pride of lions at a giraffe kill. To get to their meal, the cubs actually disappeared inside the carcass, emerging with their round little faces covered in blood. But they were wary of their father as the male always demands first choice of the food and only after he has gorged himself does he retire and allow the rest of the family to finish the remains. In the late afternoon a clan of spotted hyaenas took over the kill and from a distance and in dead silence, we watched and listened to them squabbling and growling over the food and their eerie, cackling, giggling sounds broke the stillness of the evening and drifted over the quiet valley. The carcass was evidently close to their den, as a number of cubs milled around with the parents. As these youngsters are totally dependent on their mother's milk until the age of approximately five months, they did not join in the mêlée. Soon night descended so we left the hyaenas to finish their meal in peace and returned to camp to enjoy a cold beer and discuss the day's viewings with the other guests, again seated round the campfire.

The days passed quickly and it was soon time to return home. We left Londolozi early, Denny and I getting back to the farm in the mid afternoon. On our arrival, a concerned Godfrey told us that the caracal kitten had died the previous night and its mate, the cheetah cub, had seemingly gone down with the same killer disease. The cub would not eat, vomited and was running a very high temperature. To save time I took the cheetah cub straight to Dave's house.

After a quick examination Dave said quietly, 'I'm afraid, Ann, it's infectious feline enteritis – cat flu.'

'How can that be, Dave?' I asked. 'We've been very conscientious about injecting the cheetahs every year with the vaccine – you know that.'

'Yes,' said Dave, 'but what we've used is "dead" vaccine and it's not always reliable. I was advised to use it because in the past cheetahs have reacted badly to the live vaccine.'

'What can we do?' I asked.

'As cat flu is prevalent among domestic cats,' he said, 'there are several research laboratories working on the problem. I know that there's a new and improved live vaccine on the market. I'll see what I can do about getting more information on it and will let you know if it's possible to use it on the cheetahs. Meanwhile, you'd better return to the farm and leave the cub with me – it's going to need constant attention.'

'What are the chances that the infection will spread?' I asked.

'Very great, I'm afraid,' Dave answered. 'The incubation period is 10 to 14 days. Just keep a close watch on the others and let me know immediately if any refuse food or start vomiting.'

I left the cub with Dave and returned to De Wildt with a heavy heart, wondering whether I would ever see the little animal again. I was comforted, at least, by the knowledge that the sick cub was in good hands. It was an agitated voice that telephoned us late the following evening to say that the inevitable had happened . . . the cub had died. Unquestionably the virus was infectious feline enteritis and Dave said he would be out the following day to inject all the cheetahs over 12 weeks of age with the new live vaccine that he had speedily obtained. He had gone into the matter intensively and was confident that the vaccine would be quite safe. That evening, Godfrey and I racked our brains wondering where the virus could have originated for we knew that in the house there were no other cats that could have transmitted it.

I slept unusually fitfully that night and on one occasion was woken by a strange crunching noise outside my bedroom window which overlooked the cubs' camp. I peered out. In the moonlight I saw a feral cat devouring the cubs' leftover pieces of food from the previous day. Here, I thought, must be the culprit that had probably brought in the infection! Early the following morning I discussed the issue with Godfrey and Dave and we decided that in future all the cheetahs would have to be kept within the 40-hectare camp that we had originally fenced. As the house did not fall within this area, special precautions would have to be taken there. Our immediate problem, however, was to vaccinate all our charges and to keep a close watch on the remaining youngsters and hope that they had not contracted the disease.

A few days later Mel started vomiting, and then Erich followed with the same symptoms. Dave took them both back to his Pretoria home where they were kept by day in a pen in his garden and brought indoors at night. I visited them daily and although they were fed by drip their condition deteriorated rapidly. After two days Erich died but Mel seemed stronger, and slowly recovered.

Devastated by the seeming pointlessness of all our striving, I felt close to abandoning the whole project. But I realized at the same time that we were by then too deeply involved in the work to give it up and that the information we had gathered was too important to allow us to discontinue our efforts. The two weeks following the inoculations were very tense as we waited anxiously for any reactions that the cheetahs might have to the live vaccine, but luckily there were no side effects. Lady, once again, did not let us down: she and her four cubs sailed through this troubled period without any mishaps.

One Saturday, about a week later, a local furrier contacted us. 'Godfrey,' he said, 'I need your advice. A farmer who lives some 40 kilometres north-west of your farm, beyond Brits, has brought me a large full-grown male leopard. The animal has been killing his livestock and he set a trap to catch it. The cat's in excellent condition and I have been offered a very good price for the skin. But I am loath to shoot it – it's such a beautiful animal! At the moment I've got the leopard in a large trap made of very strong welded mesh. Can you help?' he asked. 'I thought that if you could hold it over in one of your camps until Monday I would contact Nature Conservation officials and ask them what they suggest. I'd like it to be released in a protected area if possible.'

'Holding the leopard for a few days should be no difficulty,' Godfrey replied, 'as long as I inform the zoo that it's on the farm. I'll see if they can offer further assistance.'

'And there's another problem,' the furrier went on. 'The farmer wants compensation for the loss of his livestock – but I think I might be able to get a business firm to sponsor the project.'

'That's fine,' said Godfrey. 'Then I'll leave that side of things to you.'

A spitting, very angry and defensive leopard arrived in a crate about an hour later and was placed – still in its cage – under the Big Tree. To be on the safe side we had decided to keep the animal in the trap, but we bolted on two extensions to the cage to enable the big cat to move more freely. We placed a canvas cover over one end to provide a darkened area in which it could hide. Dealing with a fully grown wild leopard was completely different from the management of the normally non-aggressive cheetah. The leopard had bruised its nose badly against the bars of the trap and was patently very unhappy about being in captivity. We left it alone and only visited it to provide it with food. As we had been warned that the caged leopard's growling during the night could attract other leopards from the mountain, and remembering the episode of a few months earlier, we entered the camp very gingerly, always first searching carefully in the overhanging branches for possible visitors.

The Press gave great publicity to our predicament as we clearly had no home for the magnificent specimen. It was with great relief and gratitude that we received a cheque to cover incurred expenses, money dontated by the Nissan Motor Company which readily gave its blessing to our project and made it possible for the leopard to be released. The Natal Parks Board was happy to accept the animal, and a private airline, Comair, volunteered to fly the cat down to a new home in Natal – Mkuze Game Reserve.

The afternoon before the flight to Natal, Dave sedated the leopard and earmarked it with a numbered tag so that it could be identified in time to come should this be necessary. While it was under sedation, Dave removed it from the trap and only then did I realize what a fine specimen it was. Much stockier and more muscular than a cheetah, its whole body was the essence of strength and power. The paws were enormous compared to the small dog-like paws of the cheetah. But it was the huge green eyes which attracted me the most – they were hard and cruel, so different from the warm, soft brown eyes of the large cats to which I was accustomed. The teeth and jaws were tremendously powerful, strong enough to hoist prey into the high branches of a tree, away from scavengers – as the leopard is wont to do. The spots on its coat were arranged in rosettes and the soft sleek fur shone in the bright sunlight. It gave me a great thrill to be able to stroke and feel the powerful muscles, relaxed now that it was under sedation. This was surely one of the most handsome of cat species and certainly an animal to be treated with respect.

While Godfrey helped Dave with the earmarking, Denny and I sprayed the leopard with an insecticide to kill any existing parasites and to prevent skin infections. All went smoothly until I sprayed its head, whereupon its ears started twitching. There followed a frenzied move all round to complete the operation and a silent sigh of relief when the cat was well and truly enclosed in a smaller and lighter crate, ready for the journey south.

Early the following morning, Dave accompanied the cat on the flight. On their arrival at Mkuze Game Reserve in the northern part of Natal, he and the leopard were taken by game rangers to a selected spot near a waterhole. With the opening of the crate turned towards the rear of the transporting Land Rover, a rope was attached to its sliding door. This rope was designed to be pulled from the safety of the cab of the truck. There was a hitch, however, and the door jammed half-way up and would not budge. Dave got out of the Land Rover to release it and just as he was heaving himself over the side of the truck, the leopard pushed its head through the half-open door, saw Dave, and lunged. Dave admits to this day that he does not know how he got back into the cab so rapidly. The freed leopard brushed past at a great speed and hit its head on the open door of the vehicle. Slightly dazed, it ran off into the bush. It was seen subsequently on a number of occasions by the park's game rangers who reported that it had adapted well to the new habitat.

With a farewell to winter, we experienced a final cold snap and now, well into September, the warmth of an early spell of summer was upon us. The next cheetah-breeding season was a bare two months away and an ambitious building programme still had to be completed. Godfrey and I planned a small hospital, covering an area of about 150 square metres, to house both a surgery for Dave and five wards with under-floor heating for abandoned cubs or sick adult chee-tahs. Attached to these quarters there was to be a self-contained flat where I could stay when it became necessary to have someone constantly on the spot. We also decided to erect near the hospital smaller maternity camps for the pregnant females, so that we could easily keep them under surveillance. In each enclosure a triangular grass hut about four square metres in size would be built as we had found that both Lady and Purry had made considerable use of theirs during the last breeding season. A thick fence of dried grass separated the camps and gave privacy and protection from cold winds. It was all a frantic rush but somehow we completed the maternity section in time and the expectant mothers had settled in nicely before the first litter of cubs arrived. Of the seven pregnant females, only one – Miriam – would be having her first litter.

In one of the camps in the Monastery we placed two males that had been caught preying on livestock on a cattle farm in the Northern Transvaal. The semen tests on the males, carried out regularly by Dave, Brough Coubrough and Henk Bertschinger, indicated very little variation in sperm-count levels. Naturally only the most fertile of the males were selected for breeding. Yet again I felt sorry for poor Fanus: he did so fancy the girls, but with such a low sperm count, he was not going to be allowed to go courting. Nevertheless, he was a great help in that he was always the first to alert us when a female was in season by paying her constant attention.

Mannetjies persisted with certain antics. He would deliberately hide in the bush and then suddenly rush out as I passed. I knew that it was only a mock attack as he would then merely stamp his forefeet and spit in a most ungentlemanly way. Old Danie would watch all this with quiet, almost bored indifference; he would take his evening meal and move away from the rest of the group and eat alone. And Mof continued to be aggressive towards everyone – cheetah or human. How well I got to know them all! I loved those early mornings and late afternoons during the mating season when I was with them. After the season it quite saddened me when Joel and I had to entice the males with food, back to the Monastery. And, as if sensing that the mating period was over, the males were always reluctant to follow us.

While the females were in the maternity camps, all work on the building of the hospital stopped for we felt the noise would be unsettling for them. Determined not to be caught unprepared again, Godfrey hired a caravan to house any abandoned cubs. It was placed within the perimeter fence well away from the reach of stray domestic cats, thus making it unnecessary to bring any orphan cheetahs to the house. The birth and mortality rate of cubs born early in 1976 was not ideal, but certainly an improvement on the previous year's attempts. The females produced 34 cubs of which 20 survived – a 58 per cent survival rate. Lady again came out tops, proudly presenting us with a litter of eight – a record for a captive cheetah. She raised her brood herself, losing only one little runt. Jill gave birth to five and raised the three survivors after two had died. Miriam produced five, losing none. Dawn had a prolapse which resulted in three dead cubs born prematurely. Kromstert, Helga and Purry abandoned their young which numbered 13 in all; of these, five survived, all of them needing to be hand-raised in the caravan. I wished then that the hospital had been completed – the caravan was bitterly cold during the night and it was a constant battle to keep the cubs warm with the aid of gas heaters. Every evening I left my cosy house and, armed with hot-water bottles and flasks of warm milk, made my way in the cold night air to the mobile nursery. I was truly pleased when at last the cubs no longer required heating or midnight feeds and I could say goodbye to my temporary abode.

Three months later, after all the mothers had brought their cubs out of the grass huts, the building of the hospital was continued. Electricity was connected, heated floors were installed indoors, and sunning pens, approximately six square metres in size, were erected outside and adjoining the wards. We were now confidently ready for the third year's breeding season.

CHAPTER 8

Godfrey

*I*n May 1976 Godfrey and I had been invited to join a group of friends on a trip to the Okavango Swamps. I was thrilled at the idea but Godfrey, although keenly interested in all aspects of wildlife, explained that as he had not been feeling well lately he did not fancy the thought of driving hundreds of kilometres in a Land Rover. Vi shared his feelings about long journeys in rough terrain so Godfrey suggested that Denny should join the group. Vi offered to look after the cubs and this I readily agreed to, knowing that Dave would always be on hand if there were any problems. And so, assured that I was leaving my family in expert hands, I left the farm quite happily. Once loaded with provisions and fuel, the six of us travelled in three Land Rovers in convoy and reached Moremi Game Reserve in Botswana three days later. On entering the southern gate, we were advised to take a detour as the main road was under water. A few kilometres further on we came upon two lions at a buffalo kill but could not stay to watch this initial sighting of game as we had some distance to cover before reaching our campsite on the banks of the Okavango River.

On our second day in the reserve, we settled in at dusk under a large canopy provided by sycamore fig, knob-thorn and leadwood trees and, weary after the long day's tracking, soon dozed off in our sleeping-bags under a clear, star-filled and moonlit sky. I was awakened later by a strange noise and, after cautious investigation, made out the huge shape of a hippopotamus grazing in the papyrus reeds below where I lay. It was strange, I thought, how close wild animals will encroach when they sense no danger or threat from man. Fortunately our fire was by then a pile of ashes – I had been told that glowing coals are to a hippo as proverbially a red rag is to a bull.

At dawn the next morning, while the water for coffee was heating on a newly kindled fire, with trowel and toilet roll in hand I climbed a steep slope away from the camp. As I reached the top I came face to face with a lioness that was sitting silently on her haunches gazing intently into the bush. If she saw me, she gave no sign of it. My heart pounded as I stood as if in a trance, only metres away from her. Those big round eyes, so alert, watched some other moving object in the bush that I could never hope to see, I thought. How strong and powerful were those firm muscles that showed through her smooth golden fur and those huge paws which with one blow could break a bone or send prey hurtling to the ground. Never once did she look my way but she surely must have known I was there. A few minutes must have passed before I came to my senses and made a quick retreat.

Back in camp, a couple of late sleepers were struggling out of their sleeping-bags, but by the time we got to the spot where I had been moments before, the lioness had gone and there was only the spoor to confirm my story.

At one of our night stops, baboons settled at sunset in the trees nearby. We were not bothered by this as these animals were truly wild and feared humans; we forgot, however, that leopards sometimes prey on baboons. One visited the troop that night making us all slide deeper into our sleeping-bags, to emerge only when its rasping saw-like growls had faded away into the distance. And so the days went by, all too quickly, with never a dull moment: elephants bathed regularly in the river nearby and hyaenas raided our larder at night. In the small boat we had taken with us we could explore the swamps by water, a wonderful way to observe the rich birdlife. We heard the unmistakable call of the fish eagle and admired the bird's graceful swoop across the water as it clutched in sharp talons a fish caught among the water-lilies. We heard the loud flapping of its wings and watched the large hops across the water of the clumsy pelican as it lifted its heavy body into the air in flight. And we saw marabou storks, like undertakers with their black and white plumage, standing silently in the shallow water waiting for an unsuspecting barbel to stray their way. These were only a few of the bird and animal species we were fortunate enough to see. That four-week trip to Botswana seemed like a venture back into the past, for this land was as untouched by man as it had been for thousands of years. In this corner of Africa there were no concrete walls or tarred roads; there was no evidence of so-called 'progress'. Here we were privileged to experience raw beauty before it could be changed. My happiness was complete and I longed to tell Godfrey of my experiences.

But on returning home I received a nasty shock – Godfrey had been seriously ill while I had been away. In the past he had occasionally complained of stomach pain but had refused to see a doctor, waving my concern away by saying that a good night's sleep would cure it. This time, apparently, the pain had been so severe that he had willingly gone to hospital for treatment. Although he laughed it all off when telling me about it, I was extremely worried about his condition. He had lost a good deal of weight and tired very easily, but he assured me that he was feeling much better and as long as he watched his diet, he would recover very soon. Seeing the distraught look on my face, he told me to cheer up and suggested that we should go together to see the cheetahs.

What a warm welcome I received from Cornell, Stephen and Di! Purring loudly, they practically knocked me off my feet. They licked my hands and occasionally gave me a quick naughty but affectionate nip. Even after a month's absence, they certainly had not forgotten me. The younger group that Vi had looked after so well was full of life but more interested in food than in my home-coming. Although it had been a trip that I had enjoyed immensely, it was good to be home. In my absence, Miriam had given birth to five cubs and I was thrilled to hear that she had reared them herself. She had become quite aggressive, allowing no one near her brood. For a first litter this was excellent.

With the addition of this new group of cubs we had to consider building more enclosures. Later known as the 'teenage' camps they were to be larger than the others to enable the youngsters to run more freely. Denny laid piping and built water troughs in these camps and while he was doing so it was quite usual to see the cubs take off triumphantly with a plastic socket or pipe whenever his back was turned. On many occasions I laughed at Denny as he chased a youngster who was dragging a plastic pipe in its mouth around the camp. The cub, both curious and inquisitive, was determined not to give up until it had investigated its prize.

Early one morning as I approached the maternity camps, I heard a cub in distress – by now I was able to recognize easily the different calls of the cheetahs. The distress cry, which was a high-pitched series of chirps together with a short low growl of anger, was located and I found one of Miriam's youngsters caught in the fork of a small tree. It had obviously tried to climb the sapling, slipped and had became wedged in the fork. Miriam was very agitated but I managed to reach the cub and gently released it, only to find that its hindquarters were completely paralysed. The limp backlegs were wet, for Miriam must have tried to pull her cub out of the groove with her mouth, but in doing so had wedged it even deeper into the fork. Fortunately it was a Wednesday and we were expecting Dave shortly. On examination of the cub, Dave found no bones broken so there was nothing we could do but to isolate and confine the hapless animal, keeping it as quiet as possible in a small area within Miriam's camp. I comforted myself with the thought that this type of mishap must also occur in the wild and that the confinement of the animals played no role in the accident. Cubs are eager to climb trees at an early age but are not as sure or as steady on their feet as other cats. As the days went by, I watched helplessly as the little cub dragged its hindquarters – it seemed injured for life. However it continued to eat well, and slowly but surely strength returned to the weakened muscles of its hindlegs. After a few weeks there was no sign of an injury and the youngster could return to its mother and litter-mates, none the worse for its tree-climbing escapade.

The post of veterinary surgeon at the National Zoological Gardens in Pretoria had fallen vacant and after serious consideration Dave had decided to apply for the position although this would mean having to give up his private practice. When his application was accepted and he was appointed to the post Godfrey and I were thrilled. Although Dave had given as much help as he possibly could in the past, the animals would now automatically fall under his care and he would not be so constantly pressed for time. Also he would now have the opportunity to undertake further research into the physiology and diseases of cheetahs and this was yet another major step in the destiny of the De Wildt Cheetah Research Centre.

Early that spring we were busy sprucing up the farm in anticipation of overseas visitors – 20 dignitaries from the Middle East. The Friday before the visit, I returned home after feeding the cheetahs to find Godfrey sprinkling chlorine into the swimming pool which lay directly in front of our house.

'We want it crystal-clear and really sparkling for Sunday,' he remarked.

We had an early supper that night and then immediately retired to bed as Godfrey felt weary and we knew we both had several tiring days ahead of us. The following morning as I switched on the kettle for coffee I heard Godfrey calling faintly to me. On opening his bedroom door I found him doubled up in pain and in a cold sweat. He told me softly that he had awakened with severe stomach pain in the middle of the night and, unable to move, had called to me several times – which sadly I had failed to hear. He was deathly pale; beads of perspiration covered his forehead and his hands were very cold. Seeing him in this condition I was stunned; I felt unable to think clearly and stood stupidly rooted to the floor. This could not be – I was dreaming. His soft voice brought me back to my senses: 'Please take me to the hospital in my car – I don't want an ambulance.'

This was real – no dream at all – and I had to act quickly. I telephoned our doctor in Pretoria and told him of the sudden deterioration in Godfrey's condition. Then Godfrey was gently eased onto the back seat of his car and I raced to one of the hospitals in Pretoria. The doctor was waiting for us on our arrival and, taking one look at Godfrey, he had him transferred to a stretcher and hurried through to a ward. What followed was a nightmare. A haemorrhaging duodenal ulcer was diagnosed and at first Godfrey responded to treatment. With a smile he told me the pain had eased and that I should not worry any longer. He begged me to return to the farm to receive the expected visitors, and I agreed. But unknown to him I remained outside his ward, for I had already arranged with Frank, Dave, Miriam, Denny and Vi that the official function would proceed as planned, but without me.

That afternoon Godfrey took a turn for the worse. He was bleeding internally and the doctors felt that he was in too weak a state to allow them to operate. Conscious all the while, he repeatedly asked me about the farm. When I bade him goodnight and a good sleep that Sunday evening he merely nodded his head and said, 'I'm so tired.'

Those were the last words he said to me – he passed away in his sleep in the early hours of the morning of 27 September 1976.

It could not be, I kept telling myself. It was not so . . . I would return to the farm and he would be there to greet me with his cheery smile as usual – it would all be a dream. As my brother, he had been my closest friend, my pillar of strength, someone to lean on, someone to solve my problems calmly and sensibly. And now, overnight, my whole world collapsed in pieces about me. Godfrey had had no enemies and had always gone out of his way to help a stranger or friend in need. He had been loved by all who knew him and I knew that he would be sorely missed by many.

If it had not been for the kindness and thoughtfulness of my family and friends I do not think I could have endured the following weeks. Although at first I wanted to sell the farm and get away from all that reminded me of Godfrey, I soon realized that this would not have been his wish and that I could not abandon or give up that which we had worked for and built up together. At the same time I felt that I could not handle it alone. My family provided the solution: Reg, my younger brother, offered to settle permanently on the farm with his wife and two children and to take over the poultry section. This would allow me to keep the citrus as a farming project and devote the rest of my time to the cheetahs. Godfrey's death seemed to draw me closer to the cheetahs and all my efforts went into the project from then on, so that in the end it became a full-time occupation. With everyone's encouragement I found a reason to continue. In time I fretted less about the past and began instead to pursue the many dreams Godfrey had talked of.

The hospital was complete by now and ready for the approaching breeding season and so I moved into the tiny flatlet adjoining the five wards and surgery. Once again I spent the early mornings and late afternoons at the female camps. New males had been tested for their fertility and the best of these joined the breeding groups. Of the eight possibly pregnant females, I had observed five of the matings, but I had learnt that this did not necessarily mean that they would produce cubs. Approximately six weeks before the litters were due, the mothers were walked down to the maternity camps thus giving them time to settle into their new surroundings. At Dave's suggestion four huts in the maternity camps had under-floor heating installed, for he felt that as the cubs were born in the winter, slight warmth under the grass bedding in the hut would encourage the mothers to stay with their young. Alternatively, if they should decide to abandon their litters, at least the cubs would be kept warm until they could be rescued. This turned out to be a great improvement: both Helga and Vi who had been bad mothers in the past now reared their young themselves and I had only two cubs to hand-raise that year.

Kromstert let us down again: she gave birth to five offspring in the open and allowed them to suckle for a day. Then she gave up and lay under a bush some distance away from her brood, ignoring their shrill calls. We removed the cubs and attempted to give three to Kromstert's sister, Lady, as she had given birth to three of her own the same day. It was with bated breath that Dave and I placed the trio, smeared with Lady's faeces, one at a time in her enclosure. She would either accept them as her own, or with one crunch of her jaws destroy them.

Again Lady proved to be aptly named. Gently she picked each one up by the scruff of its neck and placed it with her own, showing her good motherly instincts. Then, without any hesitation, she licked the cubs clean and allowed the hungry youngsters to suckle from her. In all we had 19 cubs that year and only two losses.

It was a great asset having these smaller maternity camps close to the hospital for at a glance we could see if all was well. As the pregnant females were given milk once daily as well as an extra feed of meat, the proximity to base again assisted us. While the mothers were all housed in the maternity enclosures, Denny increased the number of female breeding camps, giving us a total of 16. We were now planning ahead for the second generation as Cornell and Di would soon be three years old. Although the approximate age of sexual maturity is two years, we had decided to wait an extra year before these young females were brought into the breeding programme.

Winter approached; the grass slowly turned brown and the fear of fire was strongly in our thoughts. Earlier in the season we had acquired from the zoo three white Swiss goats to assess how much grazing they could do and how many goats would be required to crop the grass short and so reduce the fire hazard. We released them into No-man's-land which held no cheetahs and where it was safe for them to wander. The area was large, 20 hectares in size, and we thought the scent of the cheetahs in the adjoining camps would keep the goats away from the fences. Often on our feeding trips we would see the three white specks on the top of the koppie. At times they wandered down the slope on the hospital side above the teenagers, but as soon as they sensed the closeness of predators they made off hurriedly, snorting furiously.

Early one morning I was awakened by grunting sounds which lasted for a short period. Soon all was quiet again and I fell asleep thinking nothing more of it. However, later that day, when feeding time came round, two young cheetahs – Harry and Desmond – were missing. A search party went out and it was not long before we found the two – next to three goat carcasses. The youngsters, little more than two years old, had climbed over the two-metre diamond-mesh fence and, having killed all three goats, had gorged themselves. We found them lying fully sated under a tree nearby. I had mixed feelings about the adventure: on one hand I was sorry to see the animals dead as they had become part of the family, but on the other hand I was thrilled that our young cheetahs which had never had any schooling in the art of killing, had caught their own prey. One aspect worried me though – why had they killed all three? Were cheetahs wanton killers? This was clearly something we would have to resolve during the next stage of our project – the release of the captive-born cheetahs into the wild. As our breeding programme was by now considered successful, we hoped to take the project a step further and to re-establish the cheetahs in suitable habitats. A number of conservation-minded people were pessimistic about this idea, and we came up against a great deal of opposition. There was a feeling that our cheetahs, although not hand-raised, were accustomed to humans, had received their food regularly and had never been taught by their mothers to kill. It was thought that because of this they would never survive in the bush. The episode with the goats was therefore very encouraging but at the same time it raised new doubts in our minds.

CHAPTER 9

Wild Dogs

A few years before Godfrey's death we had decided to buy an adjoining farm that had come onto the market. Fifty hectares in extent, three-quarters of the land was covered with virgin bush and the property extended to the top of the Magaliesberg range. At the time, we had felt that should we want to expand one day, the flat grassy plain could be fenced and would be an ideal extension to our cheetah area. But many other features also appealed to us: the land had never been cultivated and was covered with indigenous bush and grass. During the rainy season the water from an underground spring flowed in a meandering stream through the farm to a large man-made dam which attracted a number of water birds during the breeding season. We thought of stocking the area with impalas and using it as a training ground to ensure that the cheetahs knew how to kill before releasing them into natural areas. I did not greatly relish the idea, but as we wished a certain percentage of our captive-born felids to return to the wild it was at the time the only practical solution. With this in mind we started to fence off the land and at the same time to look for impalas for the project.

Because of the considerable cost and the large area of ground required for wildlife, it is quite a common occurrence for enthusiasts to join forces and to obtain a large tract of land jointly, to share the expenses of its upkeep and to fence only the outer perimeter. Such was the case at Swartkloof, near Warmbaths, 166 kilometres north of Pretoria, where two of the co-owners were Hennie and Koba

(25) Miriam and cub.

(26) Lady's cubs would lick the fur around her mouth, attracted by the scent of her recently consumed meal of meat.

(27) Three-month-old cubs clearly showing the grey mantle and dark underbody.

LEFT AND ABOVE: *(28 & 29) Cornell at the age of four months.*

(30) Erich with his 'adoptive' mother, Girlie.

ABOVE AND BELOW: *(32 & 33) Cheetahs are afflicted naturally with external parasites and they are accordingly dipped and sprayed regularly.*

PREVIOUS SPREAD:
(31) Males in the monastery.

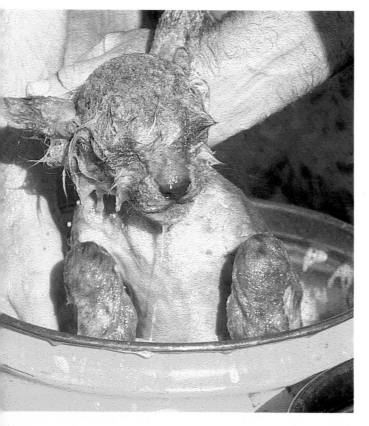

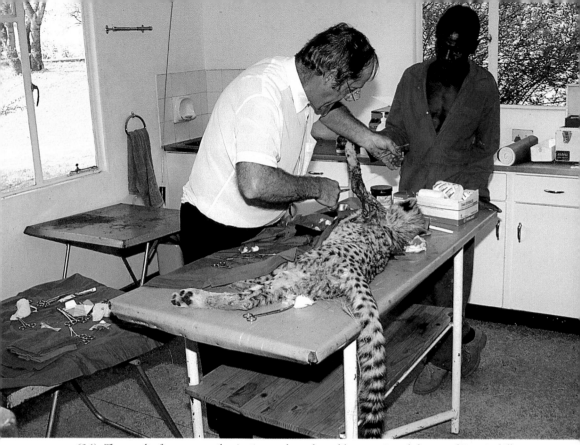

(34) *The result of an over-enthusiastic rough-and-tumble necessitated the use of Dave's skills.*

(35) *Frik was sedated and taken to the hospital; a hat was used to shade his eyes.*

(36) The tail of the cheetah helps with balance.

Diederiks. They had for some time shown great interest in our project and helped whenever possible.

One morning, while discussing developments with Koba, I mentioned our cub-training programme and asked whether there were any surplus impalas available at Swartkloof.

'I know that because of overpopulation of game in the reserve, culling is now in progress,' she offered. 'I'll certainly ask the co-owners whether they'd be willing to donate live animals to De Wildt. Give me a few days to discuss the matter with them and then I'll come back to you with an answer.'

The reply was positive and each of the 19 owners donated one live impala to our scheme. But unfortunately the bush at Swartkloof was so dense that it made it impossible to catch or dart the antelope. Added to this, because of a shortage of diamond-mesh fencing, our camp on the farm had not yet been sealed off, and so we reluctantly shelved the idea. Looking back, though, I realize that our proposed enclosure was not nearly large enough to accommodate both cheetahs and impalas. In such a confined area the predator would simply have chased its prey into a fence thereby breaking its neck – and this was certainly not the killing technique we wanted the cheetahs to use.

Dave and I decided instead to attempt a release of three of our captive-born animals into natural wild conditions where, with the help of radio-tracking collars, we would soon know what chances they had of survival. First, however, a suitable area had to be found and permission and permits obtained from nature conservation bodies and, as this would take some time to arrange, we programmed this project for two years ahead. Although we had given up the idea of attempting to train the cheetah cubs to hunt, I still wanted to have a few impalas in the area, and was therefore extremely grateful when the Natal Parks Board offered me 20 of the antelope in return for the leopard we had sent them. Often taken for granted because of its proliferation as a species, the impala is one of southern Africa's many attractive and graceful antelope. Found in large numbers and only in Africa, it is of medium size and has a reddish-brown coat and white underbelly; the male carries horns – a gracefully lyre-shaped pair that are sharp and ridged. Impalas fall prey to a number of predators but when the animal is startled its long slender legs enable it to leap swiftly and soar gracefully through the air, covering up to 10 metres in one spring, thus often confusing its attacker. During the rutting season, adult impala males fight for dominance and territory, and the successful suitor takes over the herd of females and juveniles, while the other males join up and form bachelor groups. It is a vigorous antelope and is easy to breed, the female usually producing one lamb annually. My small herd of 20 flourished and very soon gave me the thrill of seeing newborn lambs directly after birth. I was pleased that we had given up the idea to use them as prey and today they and their offspring still inhabit the area – as they must have done naturally many years ago.

At feeding time one day, Fanus refused his food and would not even bother to smell it. Dave was called and, after taking a blood sample from Fanus, he returned to Pretoria only to be back a couple of hours later with the news that the cheetah was suffering from a tick-borne disease of the blood caused by haemobartonella. It is a slow, insidious illness that destroys the red blood cells of its victim. Dave administered an antibiotic to counteract the infection and for a week afterwards, fought to save the life of Fanus, every day replacing fluid in his dehydrated body by means of a drip. Drugs were given to fight the disease and help the anaemia . . . but it was as if dear Fanus had given up the will to live. He refused to eat and his body became emaciated, his eyes sunken. A lump came to my throat as I battled to hold back my tears on the day that he died. He had always been an especially lovable animal.

I was thankful that Dave had been available during this period for very soon afterwards he flew up to a farm in the Timbavati Game Reserve in the Eastern Transvaal to fetch the celebrated white lion cubs that had been presented to the Pretoria Zoo for breeding purposes. So-called 'white' lions are creamy white in colour and as they have the amber eyes common to other lions, are not albinos. Their colouring is caused by a recessive gene; carriers usually have the normal-coloured coat and only when two carriers mate is there a chance that 25 per cent of the offspring will be white. Unusual and beautiful, they had been presented to the zoo because of the fear that they would be an easy and profitable target for the trophy-hunter if left in the wild. Dave also brought back with him four cheetah cubs which had been found abandoned in the bush – it was thought that the mother had been killed by a leopard. We were pleased to receive the four as it meant new blood to our breeding stock. One of the cubs that was exceptionally friendly and obviously the leader of the group, we called 'Mr John'.

It had been some time since I had had cubs in the hospital for hand-rearing and I thoroughly enjoyed their company, especially in the evenings. While I watched television Mr John would sit on my bed next to me. He seemed intrigued by the images on the screen and became very excited during cowboy films involving horses. He would jump off the bed and try to get behind the television screen to find out where the disappearing horses had gone. The other three were not very impressed with this show; instead they preferred investigating the bathroom where, having found the end of the toilet roll, one would come into my bedroom proudly pulling the end of the paper in its mouth, draping it over any obstacle that was in the way. They invariably did this when an interesting programme was being screened, and by the time that I had freed the paper and rolled it back on-to the holder, the programme had inevitably ended. If the toilet paper was out of reach they would raid the wash-basket. After knocking it over, one cub would grab the first article of clothing that came into view and then, with the other two helping, it would triumphantly drag the garment as if it were a kill into the surgery where they attempted to devour it in peace. Needless to say, when I had rescued my clothing from three sets of sharp little teeth a great deal of mending had to be done. But I loved the look of glee in their bright eyes when they got up to these naughty pranks and, like an indulgent parent, I found it difficult to scold them.

One evening I went out to dinner and left the cubs safely, I thought, behind closed doors in one of the wards. However, on opening the front door on my return, I heard a scuffle inside and saw four small figures quickly disappearing from view. I could not believe my eyes: the mischievous little devils had opened their door by jumping up at the handle and upon discovering a carton of fresh eggs in the kitchen had decided to play soccer with the contents. Raw eggs, eggshells and whole eggs were all over the flatlet, and of course toilet paper was draped from one end of the hospital to the other. I had never realized until that evening how much paper there was in a toilet roll! As four naughty faces peered at me from their darkened room, I burst out laughing, whereupon they came running up to me, purring loudly to welcome me home. I was quite unable to be angry but from then on I made sure that their door was securely locked whenever I left them alone.

During the day the cubs had access through my bedroom window to a large enclosure outside and would jump back and forth as they wished. One afternoon, while taking a group of friends up the koppie to see the adult cheetahs, I left the hospital in the care of my sister, Margaret, who contentedly busied herself with her hobby, dressmaking. Margaret had always tended to regard any insect, reptile or animal, including the cheetah, as potentially dangerous but I assured her that she was quite safe in my bedroom. Unfortunately I forget about the open window. On my return two hours later a muffled call for help came from the bathroom as

I entered my flatlet. Friendly Mr John, who always enjoyed human company, had hopped in through the window on seeing my sister inside. Margaret had got such a fright that she had dropped her sewing and fled to the nearest room – which happened to be the bathroom, and there she had stayed until my return. My remark, 'Oh, Marg! It's only Mr John: he won't hurt you,' did not improve matters. And worse still, Mr John had discovered Margaret's abandoned sewing and now was outside showing his mates the latest fashion. Needless to say it was a somewhat bedraggled piece of material that I recovered from the cubs – and it was some time before my sister visited me again.

The weeks passed quickly and the cubs grew rapidly. I was sad when they had to be moved out into a bigger enclosure; the hospital became quiet again and I missed their companionship. Outside, a number of small camps were popping up to house various wild pets that I was continually being asked to look after. Stompie, now almost 10 years old, was still in good condition though much to the dassie's frustration Dave had to clean his teeth occasionally. Frank had supplied a tame mate for him from the zoo and the two got on very well together. They shared a camp with a pair of mongoose, both of which were tame. The fifth inhabitant of the camp was a white rabbit. When we had fed rabbits to the cheetahs this animal had escaped and had lived in a wild state on the farm. A few months later when we had managed to catch him, I had refused to allow him to be slaughtered, and so he, too, had joined this group. In addition, I had been presented with two spotted eagle owls which had been found in a highrise flat in Johannesburg and had been confiscated by conservation authorities. Having been fed the wrong diet as youngsters, their wings were now permanently deformed and as a result they were unable to fly. Their enclosure was very close to the hospital and whenever I returned late in the evening I was welcomed with a low hoot, to which I learnt to respond – much to their satisfaction, I like to think, for they always returned the greeting. Although I enjoyed living in the flatlet, I found the accommodation very cramped, especially when I had to entertain visitors, so I decided to employ a builder to erect a thatched cottage in an area close to the hospital and maternity camps. I had to accept the usual frustrations and setbacks that go with any building operation, but soon was settled in a new home.

On the scientific side, Dave, Brough and Henk were continuing with their research work and the young males of our first litters now became part of the breeding programme. Stephen and Harry were considered potential breeders and the females Cornell, Di and Jean would soon give us second-generation cubs. Semen and blood tests were done regularly, and by now Dave had changed his inoculation routine to a yearly live infectious feline enteritis vaccine injection which he administered with a blow-dart. This method certainly was a great improvement as it did not upset the cheetahs and did away with the animal stress and human effort of catching each one of them in a crush.

In addition to the cubs born at De Wildt, the family increased with the arrival of other cheetahs, of which Taga was one. This adult male, owned by Dr Eddie Young who was at the time Assistant Director of Research for the Transvaal Division of Nature and Environmental Conservation, had had quite a varied life. Before the implementation of restrictions prohibiting indigenous wild animals from being kept as pets, he had been brought up by a family in Pretoria and had accompanied his original master to the local pub on Saturday evenings. On reaching maturity, Taga was given to Eddie, then working as state veterinarian in the Kruger National Park, and there he attempted to release the tame animal into the wild. He took Taga into the bush and let him loose near a herd of impalas. At first the cheetah made no attempt to kill but merely ran into the middle of the group, sending the antelope stampeding in all directions. Eddie then had to starve Taga for a couple of days and try again. With an empty stomach, the big cat did what was expected of him and brought down his first kill: there was no doubt that he realized that the impalas were potential food.

But Eddie still felt that Taga would not survive in the wild indefinitely as his charge had become too dependent on humans and at the age of six years (almost half his expected lifespan) had not yet learnt the cunning ways of the bush. So, when Eddie left the Kruger National Park and was transferred to Pretoria, Taga joined us. He was a lovable animal who ran up and greeted us with a loud purr whenever we entered his camp. Unfortunately he showed no interest in the females and died a bachelor at the age of 13. Through experience we have found that the lifespan of cheetahs in captivity, on a fat-free diet, is usually 12 to 16 years, whereas in the wild it is estimated to be approximately 10 to 12 years. Because of their lack of speed and poor condition, aged specimens often fall prey to other predators, such as lions or leopards.

By 1978 our breeding achievements with the cheetahs had improved considerably and we now decided we were in a position to undertake further research into a second breeding project, one involving another endangered species, namely the wild dog or Cape hunting dog. For breeding purposes six pups had been donated to the zoo in Pretoria by the Department of Agriculture and Nature Conservation of Namibia, and it had been agreed to send them to De Wildt. Wild dogs are nomadic and run in packs, often moving out of protected areas and onto neighbouring farms where they attack livestock and are shot on sight by most farmers. The low hoot of a dog in distress – one that has been deliberately caught in a trap by a farmer – brings the pack running to its aid and in so doing makes the animals an easy target for the farmer's gun. This method of extermination and the fact that packs wander and do not usually remain in protected areas, has resulted in wild dogs being placed on the list of animals regarded as 'endangered' species.

The wild dog is dependent on regular supplies of water and thus it is not found in the more arid parts of southern Africa. It cannot survive in forested or mountainous areas for, like the cheetah, it needs the expanse of the plains to run down its prey. This apart, the lifestyle and habits of the wild dog are quite in contrast to those of the cheetah. Living in close-knit family packs of 15 to 30 animals, groups are controlled by the dominant male and female; the subordinate animals accept their status, and family ties are very strong. Field research has established that generally only the dominant pair mate and the offspring, usually eight to 12 in number, are raised by the whole pack to which the survival of these pups is of prime importance. All the members of the family show great affection towards the youngsters which are very obedient and submissive towards the adults.

On awakening, either in the cool early morning or late afternoon, the dogs become tense and alert and uttering a high-pitched almost hysterical twitter they excitedly nudge and lick one another's faces, tails wagging, as they gather for the hunt. Then, with salutations completed, they leave the mother behind – with a couple of the older dogs to help look after the pups in the den – and all set off together at a steady trot. Working as a team, they make no attempt to conceal themselves when out on the chase, nor do they lie in ambush or stalk their prey; but, having selected their victim, the dogs with boundless energy run down the hunted until it is completely exhausted. And when the quarry falters, they bear down on it and within minutes tear it apart with their extremely powerful jaws, gulping down large portions of meat torn from the carcass. For this reason they are looked on by many with horror and disgust. But when I think of this so- called 'gruesome' method of killing, I always remember the statement of Jane van Lawick-Goodall in the book, *Innocent Killers*: 'The victim is usually dead within a couple of minutes and undoubtedly in such a severe state of shock that it cannot feel much pain.'* I have also heard from game rangers in the field, that whereas lions have been known to take over an hour to kill their prey, wild dogs are among the fastest and most efficient of killers. Concentrating on the sick and maimed in the wild, these dogs are on this earth for a reason and who is man with his poachers and cruel methods of trapping to criticize or condemn their existence?

Having made their kill and eaten their fill, the wild dogs return to the den where they are welcomed by the eager and hungry youngsters who, yapping incessantly, scamper out to meet them. They lick the faces of the adults and push their small snouts into their mouths, encouraging them to regurgitate a certain amount of the pre-digested meat. This is then gobbled up by the pups as well as by the older dogs that had remained behind as baby-sitters.

Dave flew to Namibia to collect the six pups. As he preferred not to sedate them they were crated together and flown to Pretoria. One of the bitches damaged her leg in transit and was held back and hospitalized at the zoo. The X-rays showed a fracture and she was operated on and a pin was inserted to support the broken bone. The other five pups were released in one of the quarantine camps at De Wildt and with excited yapping, at once ran about the enclosure, seemingly happy to be free again after their confinement. They settled down and ate well, being fed on milk and raw meat with necessary vitamins and minerals added. They were then approximately 12 months old and would reach maturity a year later.

While watching them explore their new surroundings, I became intrigued by the colour pattern on their coarse coats for no two dogs were alike. The pelt is a kaleidoscope of golden-brown, black and white blotches which cover the head, body and legs while the bushy tail ends in a large pure white tuft. A distinct black line runs down the middle of a broad and usually light-coloured forehead to disappear into the black of a short, broad muzzle. The dogs stand between 60 to 70 centimetres in height; their legs are long and slender and they have only four digits on each foot. The body is streamlined, its most unusual anatomical feature being large and rounded ears. All of this perhaps explains the wild dog's scientific name, *Lycaon pictus*, which is a combination of Greek and Latin words meaning 'ornamental wolf'.

While I studied my new charges, my thoughts turned to Godfrey and I wished so much that he could have been with me to admire and enjoy these dogs, to plan in his practical and methodical way the building of new enclosures, and to share the challenge of this new and exciting venture.

When the female with the now-mended leg was reunited with her brothers and sisters, the welcome she was given was quite extraordinary. Recognizing her at once, they howled in greeting, and she in return chewed frantically at the bars of her crate in an effort to get out to join them. Once released, the group ran excitedly up to her, twittering continuously while she, with tail wagging, crouched down submissively before them. They jostled, pushed, smelt and licked her and although they had been separated for six weeks, there was no doubt that they recognized her.

Soon after the arrival of the dogs, I entered their enclosure one day with two workers to clean the camps and to renew the water supply. I proceeded to pick up the bones left from the previous meal while Joel and Shironga cleaned the water trough. Suddenly I found myself surrounded by the dogs. I chased away one that approached from the front, only to find others stealthily creeping up on me from behind. On seeing my predicament, Joel and Shironga quickly came to my aid. From that time on I never entered the wild dog enclosure without a whip-stick for protection. Afterwards I surmised that the dogs had been interested in the old bones that I had held in my hands, but at the time I had no intention of finding out the reason for their aggression – all that had concerned me then was my self-preservation. It was interesting to observe later that we could safely enter a camp that contained a single dog, for on its own the animal was not nearly as aggressive as it was when it had the backing of its mates.

For breeding purposes we decided to pair the dogs off in smaller camps and four new enclosures were erected near the Big Tree. Drinking water was provided in buckets but we soon became aware of the fact that the dogs loved bathing. They not only immersed their heads in their buckets of drinking water, but scooped out the liquid with their forepaws, invariably knocking the container over soon after we had filled it. Eventually Denny built a large trough in each camp and it was rewarding to watch the dogs, with bodies completely submerged, revelling in the water. From then on, apart from a daily visit from one of us at feeding time, they were left alone.

* Hugo and Jane van Lawick-Goodall, 1970. *Innocent Killers.* London: Collins, p.13.

CHAPTER 10

Enter the Scavengers

*I*n the early days, Godfrey and I had often noticed the imprints of a combination of very large and smaller spoor in the soft sand behind the koppie. We surmised that these impressions had been made by a leopard and its cub. Later we were to learn that the tracks were those of a single hyaena, an animal whose forefeet are much larger than its hindpaws. Both brown hyaenas and jackals still live in the Magaliesberg range but, being extremely timid and nocturnal, they are very seldom seen. The brown hyaena is basically a scavenger, living on carrion and wild fruit.

Since the start of the cheetah-breeding project, it has become a common occurrence to see both hyaena and jackal spoor in the area where we dump the old bones, the latter being collected regularly from the cheetah and wild dog enclosures. When I first noticed the spoor it saddened me to think that these scavengers were forced to eat leftover dried meat and bones to eke out their existence and so to increase their supply of food we added to the heap of bones any dead hens we had from the poultry farm.

Never having seen a brown hyaena in the flesh, I was intrigued when one chilly morning during the winter of 1977 I received a telephone call from a farmer who lived about 20 kilometres westwards along the Magaliesberg, reporting that he had caught a brown hyaena. 'I baited it into a trap last night,' he told me, 'and I wonder whether the zoo would be interested in taking it over.'

'We don't have a breeding nucleus of brown hyaenas at De Wildt,' I replied, 'but we have considered the idea. Let me discuss the matter with the zoo and I'll let you know.'

Dave and I spoke to Frank and he agreed that it would be an interesting project to undertake. 'We know very little about this nocturnal and secretive animal, apart from the fact that it is a poor breeder in captivity,' he told us. 'I suggest you accept the hyaena and if my council members agree we'll start a small breeding unit at De Wildt. But remember, your usual diamond-mesh fence won't be strong enough. You'll have to erect a much sturdier enclosure.'

As hyaenas have extremely powerful jaws and teeth that enable them to crunch bones and tear apart a carcass or the remains of a kill left by predators, they would be able to simply pull apart any ordinary fencing. Their feet and claws are well developed for burrowing and they would therefore also have no problem in digging under the 30-centimetre curbing. Welded mesh was the only answer. But where would we obtain it at such short notice?

I thought of Koba Diederiks who had by now been appointed our honorary public relations officer.

'Koba?' I said a little hesitantly. 'I don't know how you'll do it, but can you get us five rolls of welded mesh for fencing our new hyaena camp?'

'Leave it to me, Ann,' she responded without hesitation. 'I think I know some-one who is interested enough in your project. Can you give me a couple of days?'

Dear, dependable Koba! I imagined her and her husband Hennie putting their heads together, deciding on a list of likely donors and then Koba sallying forth to convince someone of the need for the fencing. Sure enough, two days later she informed me that welded mesh fencing had been donated and that it was already on its way. It was decided to erect three breeding camps each of which allowed for expansion and in a position on the rocky area of the original 40-hectare camp. The fences were to be two metres in height with a strip of concrete a metre wide along the inside of the mesh.

While we were busy erecting the enclosures for the hyaenas we held the captured animal in an extended cage, just as we had done previously with the leopard. I visited it daily to give it food and water, and found the animal's reactions completely different from those of the leopard. While the leopard had furiously stormed the bars whenever I approached, the hyaena timidly retreated, occasion-ally uttering a low muffled growl. Later, when it was released into the new enclosure, I had the opportunity to study my charge more closely.

Similar only in shape and size to the spotted hyaena, the brown hyaena's body is dark brown, almost black in colour and is covered with long coarse shaggy hair. Its smooth black and grey striped legs carry strong well-developed forequarters and a sturdy neck. Its back slopes down to low and less-developed hindquarters giving it an ungainly stance and a clumsy lope when running. A thick blonde almost golden mane encircles its neck, and I noticed that the hair glistened when the sun's low rays caught it in the early morning or late afternoon. Unlike the spotted hyaena – known killers that live in clans – this quiet retiring carnivore, with bear-like head and short pointed ears, gives way to other animals during the day but becomes alert and fearless when darkness falls. It has a strong sense of smell and excellent night vision and usually prowls alone in search of food.

At the time Professor John Skinner, head of the Mammal Research Institute at the University of Pretoria, had assigned two zoologists to study the status of the brown hyaena in the wild. With his help and that of his team, from now on any hyaenas that were caught – usually because they were causing problems for farmers – were offered to us for our breeding programme if they could not be released in suitable areas. Twenty-eight brown hyaenas – 16 males and 12 females – had so far been trapped by land-owners in the Transvaal and these had all been quarantined at De Wildt. With the assistance of the Mammal Research Institute, 22 were relocated in nature reserves in the Eastern Transvaal and Northern Natal. The remaining six were integrated into the breeding programme at De Wildt. Four of the six turned out to be youngsters and we reckoned that it would be another two years before they reached sexual maturity.

A year or so later, a breeding pair of brown hyaenas and their four cubs were offered to us by the Natal Parks Board. The animals had been born and raised in captivity and although an attempt had been made by rangers to rehabilitate and release them back into a game-park, the exercise had proved unsuccessful. The semi-tame hyaenas kept returning to camp and it was felt that they would become a danger to the park's visitors. We were thrilled by the offer and were even more excited when the six arrived at De Wildt late one afternoon. When we released them they showed signs of being a little stiff from travelling, but were in good condition. The presence of these newcomers attracted wild brown hyaenas from surrounding areas in the Magaliesberg: they came to investigate and at the same time found the site where we dumped old bones left over after cheetah feeds. On occasions late at night I heard the sounds of a fierce battle being waged outside the perimeter fence and presumed that they came from encounters between resident brown hyaenas and those trespassing in search of food. During these altercations there was no sound of cackling or laughing as with the spotted hyaena, but instead a ferocious and savage outburst of maddened growls that shattered the stillness of the night. The morning after such confrontations, all that remained were tufts of blood-stained hair lying in a flattened grassy area.

While feeding my huge and ever-increasing family I had often noticed the occasional vulture circling high above the cheetah camps. It seemed to me that they were afraid, for despite the fact that they were obviously searching for food, they never settled on the ground to eat the cheetahs' leftovers. I had read of the findings of the Vulture Study Group – a body formed to investigate the plight of these birds – concerning a colony of approximately 350 rare Cape vultures which was nesting on the southern cliffs of the Magaliesberg range near our farm. I assumed that the birds I saw came from this colony and knew that their numbers were slowly diminishing. In nature, vultures usually bide their time and fly in at a kill only after the predators have had their fill. As calcium is a very necessary part of their diet, the birds after filling their crops with meat will normally consume some of the slivers of bones broken and left by hyaenas. However, in a fast-progressing farming area like ours where predators and scavengers were few, vultures were not getting sufficient or proper food. We had already received a number of young disabled birds which, during a recent survey, had been removed from their nests by members of the Vulture Study Group. These fledglings had deformed wings and would never be able to fly, so they were sent to us with the idea of starting up a breeding colony. Vultures reach maturity at the age of approximately five years and raise only one or two chicks each year. We placed the young birds in a large uncovered enclosure in which we planned to build an artificial cliff. Now, some years later, this structure has been completed and the vultures are daily seen here sunning themselves. We hope that during the next breeding season the birds will make use of our man-made rockface. If this project proves successful, the young healthy offspring will be introduced into the wild colony close by and will in a small way help to increase the numbers. I understand that vultures are gregarious and will accept into the flock foreign members of their own kind. But only time will tell whether or not our experiment will succeed.

One day, as I watched a lone bird with wings outstretched, gliding gracefully high in the sky, I wondered whether it would be possible to expand the feeding ground of the free-roaming nocturnal hyaenas and jackals to an area where the vultures would be able to descend and eat unafraid during the day. I felt sure that we could help these birds in their quest for food, even if it entailed asking for and collecting unwanted animal carcasses from neighbouring farms, or arranging to fetch reject meat and bones from the Brits abattoir, a mere 20 kilometres away. To supplement such supplies we also could provide the offcuts from the meat that arrived twice weekly from the zoo for the cheetahs.

In our butchery all fat was removed from this meat and the large three-to-four-kilogram portions selected for the cheetahs were packed in crates while the smaller pieces, together with the ribs and belly wall, were set aside for the wild dogs and hyaenas. All crates were then stored in the adjoining walk-in deepfreeze room, making for rapid stocktaking and easy removal when needed. Vitamin and mineral powder was sprinkled on the meat just before feeding. There were always offcuts and offal which could now be used daily as food for the vultures. Gradually a plan evolved and an isolated feeding area was found, far away from human interference. Once again I called on Hennie and Koba for help. They introduced me to people in the district who could supply reject meat and we established a vulture restaurant. Offal was obtained once a week and bone fragments were added to the meal. We felt we were providing a five-star menu and it was now up to the vultures to sample our varied delicacies. But, as is usual when dealing with wildlife, the expected did not happen and the birds did not come to feed. Hyaenas and jackals, under cover of night, soon discovered the new feeding ground and carried off portions of the dumped carcasses to their lairs. But why, I wondered one day while watching a couple of the large birds circling high above me, why would they not come down? How long would we have to persevere before we gained their trust? Or were we merely wasting our time?

After many frustrating months I stumbled upon the answer: the area I had selected for the restaurant was too densely bushed and so I changed the position of the feeding spot to a more open and unvegetated area. I had slowly come to realize that vultures, although so graceful in flight, were clumsy and ungainly on the ground and with a full crop after a good feed needed a long run to be able to lift their heavy bodies into the air. Any thick bush hindered ascent and the birds sensed that danger could lurk within the undergrowth. After 15 months of perseverance, soon after dumping a dead cow in the new feeding spot one day, vultures started circling in the sky. There were only a few but this was at least a beginning, I thought. Watching from a distance I waited patiently and was thrilled when the first vultures with feet down for landing, dropped like stones near the carcass. Others soon followed and although I was too concealed to observe them eating, I clearly heard the sounds of bickering and hissing that they make when fighting over food. Later that afternoon I went to investigate: there were no vultures in sight then and I found only skin and the clean white bones of the carcass from which they had rapidly torn and eaten the meat. It was hard to imagine how their sharp curved beaks had managed to strip the flesh from within the tough outer hide. 'At last,' I told Denny and Vi triumphantly, 'we have succeeded! It was worth it!'

Carcasses came in regularly and as we obtained them they were taken immediately to the dumping ground; slowly I gained the trust of the birds. Having changed our cheetah-feeding time from early evening to early morning, now as I left the butchery with the bakkie loaded with meat for the animals and a bucket of offal and offcuts for the birds, I could see vultures circling high above me. At times over 100 of them would appear, seemingly from nowhere. After first feeding all my four-legged friends I would then head for the open area. The birds would swish down low above me and in that moment I would feel a surge of tremendous joy – truly a great reward for the time I thought I had wasted. Keeping my distance, I would watch in awe as the vultures came silently in to land, feet thrust stiffly down and wings outstretched and motionless. Looked on by many as ugly birds, vultures in flight are to me birds of beauty and grace. A number of vulture restaurants have since been established on farms along the Magaliesberg range and it is interesting to note that the number of chicks sighted in 1990 had increased dramatically.

Soon other scavengers came to enjoy the restaurant offerings: marabou storks and black and tawny eagles were occasionally sighted, while yellow-billed and black kites migrating south to the warmth of summer, set their course for De Wildt where they knew there was an abundance of food – much to the disgust of the busy, noisy crows that had taken up permanent residence. I am convinced that were a count to be taken, ours would be found to be one of the largest populations – if not the largest – of pied crows in the country.

With Godfrey gone, I now had difficulty in leaving my large family whenever I needed to go on holiday. I discussed the matter with Frank and he offered to send out from the zoo a young and energetic nature conservationist, Eugene Marais, to take over the reins while I was away. Keen to help, Eugene soon knew all his charges by name and I was able to leave with the knowledge that the farm was in good hands. Eugene was extremely interested in birds, especially in raptors, and under his supervision, cages were built at De Wildt for a varied group of indigenous owls. These included the wood owl, Cape eagle owl, pearl-spotted owl, white-faced owl, marsh owl, grass owl and the common barn owl and spotted eagle owl. Most of these birds had come to us with injuries and it was distressing to realize how many of them had been hit by cars at night. A pair of Cape eagle owls was the first to arrive on the farm and to our delight soon hatched out three beautiful chicks. At first they were small balls of white fluff but as they grew older they puffed themselves out and stared angrily with large brown eyes as they tried to make themselves appear alarmingly aggressive – but to me they only succeeded in becoming more appealing.

Stray birds and animals continued to be brought to the farm. There were vultures that had flown into electricity cables, kites with broken wings and even a young marabou stork that was badly infested with lice. While Eugene attended to the wounded birds, I accepted the injured animals. Every effort was made to release these birds and animals once they had recovered, and only those that we knew would not survive in their natural environment were kept in captivity at De Wildt. Among those remaining with us was one of my biggest headaches – a female vervet monkey with a baby which had been brought to us by some local villagers.

During the dry winter months, because of a shortage of food, monkeys venture down from their mountain habitats and raid small villages and farms, scavenging whatever supplies they can find. As this particular mother monkey had a young-ster to feed she was understandably hungry; she became very aggressive and inhabitants of the raided village in desperation set a trap in the form of a large tin. It was filled with mealies and at the top had an opening just wide enough for the monkey's hand. As the villagers had anticipated, she tried to steal a handful of corn and since she refused to release the mealies when approached she was easily caught. The villagers placed mother and baby in two hastily made crates and a pair of screeching monkeys was delivered to my door. I telephoned Dave imme-diately: 'If possible, I'd like to release them on the mountain,' I said, 'but do you think the mother will accept her youngster? They've been separated since they were caught.'

'I suggest that you place them together in one of the enclosed camps – the ones outside the hospital are ideal,' said Dave. 'Then watch the mother's reaction. And there's another thing to watch – monkeys have damned sharp teeth!'

I proceeded to carry out Dave's instructions. However, by now all maternal instincts had vanished and self-survival was foremost in the monkey-mother's mind. She flew at the youngster whenever it came near her and would obviously have killed it had it remained in her cage. We recaught the mother, took her behind the koppie and released her in the thickly bushed area where we had on several occasions seen other vervet monkeys. As she was a young female we hoped that she would be accepted by the resident troop. My attention now turned to the baby which, barely two weeks old, was completely harmless and clung to me in terror. Our hospital was fully occupied at the time and, as I was feeding cubs as well, I had more than enough to do. Reg, my brother, and his wife, Judy, although very busy with the running of the poultry farm, were always available to lend a helping hand. And so, using a handkerchief to act as a nappy, and wrapping a blanket (that I had used for the cheetah cubs) around the tiny helpless creature, I took the pink, miniature human-like object to Judy and asked whether she would be able to look after it for a while. With one glance at the baby, Judy's heart melted and willingly she agreed to take it on. I heaved a sigh of relief. He was named 'Zack' and in a short time accepted Judy as his foster mother. Clinging tightly to her shoulder and chattering all the time, he would nestle his head in her hair. It was endearing to watch the tiny fingers gripping the baby's bottle at feeding time and witness the look of utter content on his little wrinkled face. But if Judy should take the bottle away from him, all hell was let loose! As he grew older and more boisterous, household ornaments and valuables had to be carefully locked away.

Like a human child, the young animal had outbreaks of temper-tantrums and had to be severely scolded. If one comes across a family of vervet monkeys in the wild it is quite common to hear the startled screams of a teenager that is being reprimanded by an adult. Zack's favourite pastime, when Reg's back was turned, was to slyly steal his box of cigarettes and, with it clutched tightly in his hands, to climb the highest tree in the garden. With an irate Reg berating him from below, Zack would slowly and nonchalantly break the cigarettes open one by one and eat the tobacco. Knowing he was defeated, Reg would return to his office mumbling monkey-curses all the way. A little later Zack would return to the house quite unconcerned, jump onto Reg's shoulder and make whispering sounds into his ear. All wrath would then disappear in an instant and Reg would have to admit that the little monkey had his master just where he wanted him.

On Sundays Reg's family usually had a lunchtime braai under the trees in their garden and on these occasions Zack was in his element. Jumping from tree to tree, he delighted in showing off in front of everyone, almost as a young human child will do, saying, 'Look, everyone, watch me!' Then suddenly and unexpectedly he would drop down onto the table below with naughty glee. I'm afraid, though, that his pranks were not always appreciated by guests. As he grew older and reached maturity, difficulties arose. He had moments of aggressiveness, usually directed towards strangers, though never towards Reg and Judy. Being a male, he saw strangers as invaders of his territory and he would feel a need to establish his dominance and manhood. We were torn between the harsh facts and our love for the little monkey who was by now a very dear member of the family. The final decision as to Zack's destiny came when he attacked Reg's 14-year-old daughter. She had always been afraid of Zack and he seemed to know this. On Dave's suggestion, he was slowly introduced to a group of monkeys at the Pretoria Zoo. Under Eugene's care he was placed in a cage adjoining the main group, enabling him and the zoo monkeys to get to know one another through the dividing fence. One by one the resident animals were introduced into Zack's cage. At first he reacted by sulking in a corner but soon one of the younger monkeys started playing with him and he slowly joined in. This episode once again proved how difficult it is to make a domestic pet of a wild animal.

In 1978 eight cheetah females had produced 29 cubs of which 22 had survived. We realized that before long we would need more land so that we could expand to meet the needs of this growing family. Frank was at the time busily engaged in a new project involving about 6 000 hectares of land near Lichtenburg in the Western Transvaal. Under lease to the Pretoria Zoo and a mere two-hour run from Pretoria, the habitat was conveniently situated and considered an ideal spot for a breeding station for many herbivora, both indigenous and exotic. The grazing was excellent there and water was abundant. To help pay running costs, the station was opened as a game-park for day visits. Now, as an extra attraction, Frank decided to fence off 200 hectares for cheetahs. Both he and Dave thought it would be a good policy to house some of the felids at Lichtenburg in case any unforeseen problems should arise at De Wildt. On looking back, we realized that we had achieved so much: in five years 130 cheetah cubs had been bred, with a birth mortality of only 37 per cent. It would be a pity to be unable to continue our work because of some unforeseen catastrophe in the future.

When the fencing at Lichtenburg had been completed, a group of about 20 of our youngsters was selected; they were to be released on the day of the official opening of the game-park. It was a great occasion for us as this was the first large consignment of captive-born cheetahs to leave the farm. We selected those that had been raised naturally by their mothers and were thus wild and assertive by nature. As Frank required at least 20, they had to be chosen from two age-groups. The animals were transported to the reserve in crates which, on arrival, were lined up ready for the cats to be released. All went according to plan and we were happy to see the cheetahs dashing out of the confining crates into the wide and spacious grassland area.

But within the first couple of weeks of the cheetahs' release at Lichtenburg there was a setback. The group of older males, some of them brothers, began to show its dominance by tackling, wounding and killing a few of the younger animals. This was a severe blow as these cats were all sub-adult and we therefore had not expected behaviour of this kind. After receiving an urgent plea for help, Dave travelled to Lichtenburg to dart and sedate the bullies and to bring them back to De Wildt. With Frank's agreement, we decided to replace the two age-groups with a smaller group consisting of yearlings that had been raised together. This proved to be a happy solution and the youngsters, now mature adults, have successfully established their territories and live in harmony in the 200-hectare enclosure. But although this was a giant stride forward, one hurdle remained: we still did not know how captive-born cheetahs would adapt in the wild.

CHAPTER 11

A Taste of Freedom

*A*nd so our next big challenge was to establish whether or not cheetahs that we had bred in captive conditions could be returned successfully to their natural environment. Just as we were considering how to set about this, a request came from the military air-force base at Waterkloof, near Pretoria, for the loan of a tame cheetah to pose for photographs with members of a unit known as the 'Cheetah Squadron'. Frank granted permission and we decided on Mr John as he was as manageable as a tame cheetah could be.

Dave arrived at De Wildt early on the morning of Mr John's great day and, led with a collar and leash, the big cat jumped without hesitation onto the back seat of the Mercedes that was to take him to Pretoria – as confidently as if he did it every day of his life. I took my place beside him and talked to him constantly, he responding with loud purrs. I wondered whether the noise of the car engine or the movement of the vehicle would upset him – but no, he sat upright and in a most lordly manner on the seat, gazing, seemingly unperturbed, out of the window and into the distance. I wished I knew what he was thinking.

Driving through the busy streets of Pretoria with a cheetah as a passenger was quite an experience. Whenever Dave stopped at an intersection, the occupants of cars drawn up alongside would casually look over towards us. On seeing the cheetah on the back seat, eyes would grow larger and larger but Mr John remained impassive and merely looked arrogantly right through the gaping humans as if this was just a routine trip for him. What did interest him, however, was every small dog that happened to be in town that day! I could feel his body tense as he steadied his gaze on the small animals and once again I wished I could read his thoughts. On arrival at the airbase, we drove through the security gates and onto the tarmac area where photographers and military top-brass were waiting. Mr John and I stayed in the car until everyone was seated and I then led him out and placed him in front of the group. Although both he and some of the men were uneasy to begin with, the tension soon eased when Mr John suddenly purred and licked the hand of one of the men. After the photographs had been taken, Mr John was escorted back to De Wildt.

While chatting to some of the officials at the airbase, we mentioned our plan to experiment with a trial release of captive-born cheetahs in the wild. Until this had been achieved, we told them, we would have no idea whether cheetahs born in captivity would survive the harsh laws of nature. We were then informed that the airforce had under its jurisdiction a game-farm at Hoedspruit in the Eastern Transvaal. On this farm there were few predators and, because the hunting of game was prohibited, the impalas were breeding too well and the farm's game rangers feared that an overpopulation of the species would lead to overgrazing and destruction of the habitat. This was exactly the type of situation that we needed for our cheetahs and we stressed our eagerness to try out the prey-catching capabilities of our youngsters. The commandant of the airbase promised to take the matter to higher command and to let us know if anything could be arranged.

A few weeks later, a meeting was held at the Pretoria Zoo. A letter had arrived stating that the airforce reserve would be made available to us so that we could carry out our experiments there for three months, after which it was intended to introduce rarer species of antelope to the game-farm. Until these species had bred successfully, predators would naturally not be wanted in the area. We felt that three months gave us ample time and, with the full support of the zoo council, we started to make arrangements for the experiment.

Three brothers, aged about two-and-a-half years, from a litter of eight that Lady had produced, were selected to be released. Named 'Rodgers', 'Gouws' and 'Jan', they were large sleek cheetahs and I fondly saw them as the 'lads of the village'. In order to follow their movements and yet not to cause them hindrance, collars with light-weight built-in radio transmitters were to be placed round their necks. Battery-charged, these transmitters, once they were activated on a selected frequency, emitted a 'beep-beep' signal which was picked up by the antennae carried by trackers. In this way the animals could be followed constantly, but at a distance. As trackers, young men with plenty of stamina and a genuine interest in the project were needed. We all realized that if the tracking was done half-heartedly and the results proved to be inaccurate or inconclusive, any future plans we had of releasing captive-born cheetahs into natural surroundings would be difficult to motivate. This was to be a significant trial release and the results were going to be very important, not only to us but also to the future conservation of the cheetah species.

We approached Dr Eddie Young for his help and advice. Keenly interested in cheetahs, he had supported our project for a long time and now he suggested that we should call in the services of Howard Pettifer. Howard, a field officer in the same department as Eddie, had been studying the cheetah population at the Suikerbosrand Reserve, near Heidelberg in the Transvaal. This was, in 1975, a fairly new game reserve of 13 400 hectares into which eight of these cats had originally been introduced but within two years had increased to 24 in number. Carnivores such as lions, leopards and spotted hyaenas are known to prey on cheetahs – especially the cubs – and the lack of these predators in this reserve accounted for the low mortality among the cheetahs. Howard's research project at Suikerbosrand was to ascertain whether the introduction of cheetahs would control the numbers of small and medium-sized ungulates, in particular the blesbok. The results of this introduction had been dramatic. As the cheetahs showed a preference for the female and juvenile blesbok and went for a fresh kill whenever hungry, the antelope population had dropped quickly and drastically, so causing great concern. Accordingly, plans had to be made to translocate most of the cheetahs to other larger reserves; for Howard's research indicated that unless the cheetah population was controlled and carefully managed in the smaller game parks such as the Suikerbosrand, the effect on antelope populations could be most detrimental.*

Howard supported our theory that captive-born cheetahs could adapt to life in the wild and, with his previous experience, we were confident that he was the ideal man to head the team that would follow our youngsters. Observing them would involve tracking the animals continuously, day and night, as we needed a complete record of their movements during the release. It was impossible for a single man to attempt this task so Howard selected as assistants Piet Muller and

Koos de Wet who had been his field technicians at Suikerbosrand. They planned to be on eight-hourly shifts, following the trio.

The terrain at Hoedspruit is typical Eastern Transvaal bushveld, an undulating landscape varying from open grassland to dense bush. With a summer rainfall, the climate of the region is very similar to that of De Wildt – long hot summer days and a short moderately cold winter. Suikerkop, the reserve in which the cheetahs were to be released, is completely game-fenced and covers an area of 1 100 hectares. The Transvaal Division of Nature and Environmental Conservation made two four-wheel-drive vehicles available to Howard and when all formalities were completed the great day arrived.

First, the three big cats were sedated, collared, placed together in one crate and freighted by air to the Hoedspruit air-force base. Howard had told us of his intention to hold the cheetahs initially in a small quarantine camp to enable them to settle down and to become acquainted with their new surroundings. We agreed with this proposal and I imagined how dramatic it must be for a wild animal to be taken from its environment, suddenly enclosed in a dark crate and then released hours later into an entirely new and strange habitat. Howard promised to let us know of their progress from time to time, assuring us that he would be a silent observer who would never interfere with or help the animals should they get into a tight corner. It was obvious that we needed to know what happened to them when they were left on their own and how they adapted to every new situation that they encountered.

While in the quarantine enclosure, the cheetahs were fed the carcasses of whole impalas that had been shot specially for them. According to Howard, the three had difficulty at first in opening the carcasses but after the fourth one they learnt to tackle them with speed and efficiency. Now confident that the cheetahs could successfully cope with their dead prey, we duly set a day for the release of the three into the reserve. I was anxious, and troubled thoughts kept flashing through my mind. Would these youngsters, not yet three years old and having so far lived a protected life, fulfil our wildest expectations? When they felt the pangs of hunger, would they know that under the smooth hide of a live antelope in flight was the meat for which they craved? Would they be able to attack and bring down their prey without having had lessons from Lady, their mother? How would they react when confronted by larger opponents? Would they realize the size and strength of these opponents and give way, or would they attempt to fight and defend themselves if attacked? So many questions filled my head, but I realized that I would simply have to be patient for they could only be answered in time and only after painstaking observations and time-consuming research. Once again I did so dearly wish that Godfrey was with us to share the anxiety, excitement and dreams of success, for this surely was the culmination of our original aim way back in the early 1970s when we had started out on our cheetah-breeding venture.

Time unconcernedly ticked on, very slowly. At De Wildt we waited impatiently for news and at last the first report came through. I was out feeding the animals when I received a message to telephone Dave urgently. I knew immediately what it was about and my stomach turned. Rodgers, Gouws and Jan were now into their second day of freedom – had something gone wrong? I hurried back to my office telling myself all the time to think positively.

Soon Dave's calm voice came through on the telephone. 'I've got good news for you, Ann. Our cheetahs have made their first kill and successfully – a giraffe calf,' he said.

I was overjoyed and could not reply; words simply failed me.

'Ann, are you there?' Dave queried.

'Yes, I'm here, but I'm so excited I can't speak,' I blurted out, laughing.

Thereafter reports came through regularly, each one confirming that our 'lads of the village' were doing what was expected of them and in this way were proving our theory that the hunting instinct of the cheetah is inborn. According to Howard, the three were adapting well and roamed throughout the reserve. At intervals of four to five days they caught a waterbuck, a giraffe calf, a young kudu or an impala. They were not always successful at catching impalas but were improving with experience. In a formal report that Howard released later it was stated that 'the poor hunting success on impala can be chiefly ascribed to the fleety impala heading for thick bush on sight of the cheetah and the fact that the cheetah very seldom stalked impala, but chased them openly'.[†]

Table 2 (overleaf) is reproduced here from Howard's report and gives an idea of some of the daily activities of the trio after their release from the small camp. The report continued: 'From this table it can be seen that the cheetah move small distances when sated and that these distances increase as the animals become more hungry. Invariably, the movements on the day following a kill were to the nearest watering point.'[†]

The Hoedspruit game-farm, although small in size compared with many other game reserves, had been ideal for the initial release but when the trial period had

DATE	DISTANCE COVERED (km)	KILL	SEX OF PREY	AGE OF PREY	PERIOD BETWEEN LAST FEED AND NEXT KILL (days)
					Table 2 The daily distances covered, kill frequency and prey species taken by three captive-bred cheetah on the portion Suikerkop, Hoedspruit Air-Force Base (Pettifer, 'Experimental release', p. 1004)
8/10/79	8,0	Giraffe	M	Calf	—
9/10/79	2,8	(feed)			
10/10/79	6,7				
11/10/79	3,9	Kudu	F	Subadult	2
12/10/79	7,1				
13/10/79	10,2				
14/10/79	8,3				
15/10/79	1,1	Impala x2	MM	Adult, Subadult	4
16/10/79	1,4	(feed)			
17/10/79	14,4				
18/10/79	11,0				
19/10/79	9,2				
20/10/79	9,9	Duiker and	F	Adult	5
–	–	Waterbuck	M	Adult	
22/10/79	4,2				
23/10/79	5,8				
24/10/79	12,8				
25/10/79	10,8				
26/10/79	2,1	Giraffe	M	Calf	5
27/10/79	3,2				
28/10/79	14,8				
29/10/79	3,6	Impala	M	Adult	3
30/10/79	1,6				
31/10/79	11,8				
1/11/79	0,7	Impala x2	MF	Adult	3
2/11/79	0,9	(feed)			
3/11/79	2,6				
4/11/79	11,1				
5/11/79	13,1				
6/11/79	18,0				
7/11/79	*	Impala	M	Adult	5

Total distance covered: 222,3 km Average daily distance: 7,41 km
*Project terminated; distance not included in calculations

expired, Howard was unwilling to send the cheetahs home. He explained that he wanted to continue the exercise but in a larger area where he hoped he could determine whether the three males would settle in a specific place or would continue to wander as they had done at Suikerkop. He also wanted to witness their reactions when confronted by other predators. Our problem was to find a suitable game-farm that was large enough, and with an owner who was interested in our work and prepared to help us continue the project.

Recently we had gained two very good friends, Dick and Nora Reucassel. Dick is a well-known professional wildlife photographer and he offered to help us in any way he could and also in an honorary capacity by providing us with photographic expertise and material. He brought us into contact with a number of people and businesses interested in conservation and whose contributions and generosity were to prove substantial. Among these many friends introduced to us by Dick we had come to know Hans Hoheisen very well. Hans, who was a great nature-lover and wildlife conservationist, owned a large game-farm measuring 22 000 hectares in extent and bordering on the Kruger National Park in the Eastern Transvaal. His land, together with a number of other unfenced private game-farms, collectively formed the Timbavati Private Nature Reserve which covered an area of over 60 000 hectares. Adjoining Timbavati was yet another large area, similarly controlled by private land-owners: known as the 'Klaserie Nature Reserve', it measured 65 000 hectares. Hans had observed with great interest the movements of our cheetahs at Suikerkop and now, on hearing of our plans to continue the project elsewhere, he kindly agreed to the release of the three cheetahs on his property. His farm was in beautiful bushveld country where water was plentiful and game was protected and thus in abundance. Elephants, buffaloes, giraffes, wildebeest, zebras and impalas were among the many animals to be found there. Of the predators, lions and leopards were observed regularly but cheetahs only occasionally. We were thrilled with Hans's offer as conditions in the Timbavati seemed to be ideal for the continuation of Howard's work.

Rodgers, Gouws and Jan were duly transferred from Suikerkop to Kempiana, which was the name given to the section of Hans's farm where the release would take place. Once again Howard held the three for a month in a 30 by 30-metre enclosure until he was satisfied that they had adjusted to the new environment. Realizing that the cheetahs would now have few fences to hinder them and would probably wander over great distances, Howard obtained permission from the authorities in the Kruger National Park and both the Timbavati and Klaserie

Table 3 The daily distances covered, kill frequency and prey species taken by 3 captive-bred cheetah in the Timbavati and Klaserie Private Nature Reserves. (Pettifer 'Experimental release', p.1007)

DATE	DISTANCE COVERED (KM)	KILL	PERIOD BETWEEN LAST FEED AND NEXT KILL (DAYS)	DATE	DISTANCE COVERED (KM)	KILL	PERIOD BETWEEN LAST FEED AND NEXT KILL (DAYS)
28/2/80	19.0			1/4/80	6.0		
29/2/80	7.5			2/4/80	1.0		
1/3/80	17.0			3/4/80	2.0		
2/3/80	9.0			4/4/80	4.0		
3/3/80	5.0	Giraffe calf F		5/4/80	4.5		
4/3/80	—		(feed)	6/4/80	5.0		
5/3/80	4.0		4	7/4/80	4.0	Giraffe calf	8
6/3/80	8.0			8/4/80	—		(feed)
7/3/80	22.0			9/4/80	0.5		
8/3/80	15.5			10/4/80	2.5		
9/3/80	9.0			11/4/80	15.0		
10/3/80	17.0			12/4/80	4.5		
11/3/80	9.0			13/4/80	6.0		
12/3/80	11.0			14/4/80	3.0	Chickens	5
13/3/80	10.0			15/4/80	3.0		
14/3/80	3.5			16/4/80	10.0		
15/3/80	2.0	Giraffe calf M		17/4/80	5.5		
16/3/80	—		(feed)	18/4/80	9.5		
17/3/80	2.0		10	19/4/80	10.5		
18/3/80	4.5			20/4/80	3.0		
19/3/80	5.0		(feed)	21/4/80	3.0	Waterbuck yearling M	(6)11*
20/3/80	9.0			22/4/80	—		(feed)
21/3/80	7.5	(fight)		23/4/80	1.0		
22/3/80	1.5			24/4/80	0.5		
23/3/80	1.5			25/4/80	8.0	Impala calf	2
24/3/80	1.0			26/4/80	2.0	Chickens	1
25/3/80	1.0			27/4/80	3.0		
26/3/80	—			28/4/80	2.0	Kudu yearling M	(1)2*
27/3/80	1.0			29/4/80	—		(feed)
28/3/80	3.5			30/4/80	1.0		(feed)
29/3/80	1.5	Impala adult M	11	1/5/80	1.0		(feed)
30/3/80	—			2/5/80	6.0		
31/3/80	1.0			3/5/80	3.0	Terminate project	

Total distance covered: 338.5 km   Average daily distance: 5.21 km * Chickens not considered a full meal reserves to follow the animals should they move onto neighbouring farms. Once released on Kempiana, the trio proceeded to explore the new area and Howard later reported that 'Over the 65 day study period, the cheetah covered 338,4 km with an average daily distance of 5,21 km'.[†] It was interesting to note that Howard observed a marked difference in the period of time between kills in the Timbavati area compared with those at Suikerkop. 'Suikerkop, being a small area with a high density of prey animals, ensured the chances of the cheetah locating potential prey within a short period of time, whereas in the Timbavati P.N.R. and Klaserie P.N.R., the cheetah often went for two to three days without sighting potential prey. Probably an equally important factor was that the cheetah were less restricted by fences in the Timbavati and Klaserie P.N.R. and spent more time and energy exploring their new environment.'[†]

Howard also noted that 'All kills with the exception of one were effected in the early hours of the morning. All the study animals had captured and killed on their own, although Rodgers was undoubtedly the best hunter. Prey was dispatched efficiently except with the first giraffe caught, which began to recover before being finally killed. All prey was killed by the typical cheetah strangle-hold on the throat'.[†] He observed that 'the cheetah engorged themselves on large quantities of meat, the distended bellies showing for up to four days after a feed. Likewise, it has been demonstrated that cheetah are capable of going long periods without food with no obvious adverse effects'.[†]

Three weeks after the cheetahs' release, word from Howard reached us at De Wildt that the three males were scent-marking in a particular area in the Timbavati Reserve. Although there are some conflicting theories resulting from research studies done in the wild as to whether cheetahs are territorial or not, generally it is accepted that unlike the solitary female, males especially litter mates remain together for a long time. My own observation indicates a lifelong attachment and it would seem that their home-range is not fixed. In order to avoid close contact with other cheetahs and establish a spacing system, scent-marking is used to demarcate territory. This is done by the cheetahs themselves by spraying urine onto tree trunks and large stones as well as by using their hind feet to scrape soil into a mound while defecating. (Unlike most other cats, cheetahs seldom cover their faeces with soil.) According to Howard, the area they were marking was ideal and it seemed likely that the three would settle there together.

With this news of further progress, I began to wonder whether we could remove their collars after a period of time and leave them alone to establish themselves in this little corner of the Lowveld. Question after question raced through my mind and while I pondered over their future the animals themselves were deciding what it was to be: they had trespassed into the territory of three resident male cheetahs that had had no intention of sharing their domain with the strangers and had decided to fight for what was theirs. Howard reported that the ensuing struggle had lasted, on and off, for eight hours with casualties on both sides. In our team, Jan had been badly mauled and had lost the sight of one eye. He was unable to move but his two brothers stayed by his side for seven days until he had regained his strength. During this period no effort had been made to hunt for food. On the eighth day, nervous and unsettled, the trio had slowly moved on, never to return to the area where the fight had taken place.

Howard's report continued: 'The three study cheetah remained together throughout the study period and demonstrated an extremely strong group cohesion. When one of the cheetah was separated during a hunt, intensive vocalization from both sides would reunite them. Perhaps the best illustration of this strong bond was shown after the fight with wild cheetah.'[†] Also: 'Face-licking between the study animals was frequent and often accompanied by deep purring. These cheetah also frequently licked each other's wounds.'[†] This confirmed what we had earlier observed at De Wildt, namely that there was always an extremely close bond between cheetah litter-mates.

As I was very keen to see for myself how the tracking was progressing, I drove to the Eastern Transvaal with friends Hennie and Koba to spend a few days with Howard and his colleagues. When we arrived there after the 400-kilometre journey, we found that the cheetahs had moved off Hans's farm in the Timbavati Reserve and were now in the adjoining Klaserie Reserve. After locating Howard we were greeted with the sad news that Gouws had been bitten by a snake, according to a post-mortem investigation report, and had died three days previously. The snake had been a puff adder and we surmised that the encounter between cheetah and snake must have taken place during the night because at no time during the day had Howard and his colleagues noticed anything even vaguely amiss. This cast a dark shadow on our meeting, for to us it was as though we had lost a member of the family, and to Howard and his colleagues it was the death of a close friend. They told us that after the carcass of Gouws had been removed for autopsy, Rodgers and Jan had remained in the vicinity, continually calling and searching for their brother before finally moving on after three days.

I had learnt on many occasions before that I could not let my feelings run away with me. We had wanted to know how these animals would adapt to natural surroundings and snakebite was, after all, one of the many hazards of the bush, one that might have involved any free-roaming cheetahs, I told myself. A comforting fact, however, was that the remaining two cheetahs were once again beginning to scent-mark a selected area in Klaserie in which they clearly planned to settle. Joining up with Howard, we bundu-bashed and bumped our way through the veld in a Land Rover which followed the remaining two animals as they progressed through dense bush and river-edge reeds. Being off the road, we had no vehicle tracks to follow and often we were forced to make a detour, with one of the rangers following the animals on foot and always keeping in radio contact with them. Those travelling in the vehicle were then obliged to find an easier route which at times seemed all but impossible as the way was blocked by large boulders, rocky outcrops, or by dense vegetation.

It was good to be back in the bush again. Living close to nature always seems to place one's life and all its complexities into a more realistic perspective. For me, the most relaxing time of all is at night when one lies under a clear and moonlit sky and can appreciate the solitude and silence of the bush, a silence only occasionally interrupted by familiar nocturnal sounds of the wild. That night the friendly campfire flickered and crackled as Howard told us the story of the past few weeks. We all listened intently, deeply interested as he related tales of how our three had instinctively accepted the area as their homeland, and of their initially unsuccessful attempts to hunt buffaloes, wildebeest and zebras – efforts that had usually ended in them being painfully kicked.

'They were very successful when hunting giraffe calves though,' Howard explained. 'While the other prey species dashed off when they saw the cheetahs, the giraffes held their ground and in some cases approached the cats out of curiosity. But in all these cases the trio only showed interest in the calves.'

Howard continued, telling us how the cheetahs had used the same technique with each giraffe kill. 'Jan kept the adult giraffes at bay while the other two caught and brought down the selected young animal,' he said. 'It was always Rodgers who pounced on the hindquarters, embedding his dew claws into the calf's rump and then dragging it down, while Gouws went for the neck. Once the prey was on the ground, Gouws would move his grip higher up and use the typical cheetah strangulation bite to the trachea just below the jawbone. It was interesting,'

Howard remarked, 'to note that the cheetah instinctively knew it should remain behind the body of the struggling victim to escape being kicked.'

'What about other predators, Howard?' I asked. 'Surely the cheetahs have come into contact with lions?'

'Yes,' he replied, 'but not very often. They are instinctively afraid of lions and avoid them whenever possible. Their reaction to spotted hyaenas, however, was more aggressive and on one occasion the three cheetahs, while feeding on a carcass, kept seven hyaenas at bay until they had finished their meal. Yet,' Howard shrugged, 'in another encounter they allowed a single hyaena to drag off their kill.'

'That's strange,' I remarked. 'Why do you think they would allow their prey to be so easily taken?'

'They were probably sated,' Howard explained. 'As you know, cheetahs eat their fill and leave the remainder for the scavengers. They were continually followed by black-backed jackals which kept their distance and approached the carcass only when the cheetahs had finished eating.'

We chatted well into the early hours of the morning – there were so many stories to hear and so much to discuss. While we did so, the two remaining cheetahs lay together contentedly, only a short distance from us. They had recently killed and eaten an impala and we knew they would rest now until the pangs of hunger were felt once more. It amazed me how much of the hunting and killing was instinctive and did not have to be taught, either by the group or by a parent. Perfecting their killing technique was naturally achieved through experience but it appeared that they knew intuitively all the basic rules. As we talked, I realized how much had been learnt from the experiment, all of it gleaned because of the enthusiasm, dedication and sheer hard work of Howard and his team. During our conversation at the campfire Howard remarked that he was worried by the fact that the cheetahs had moved over to the Klaserie area. Although larger than the Timbavati Reserve, here each land-owner was allowed to house a family of workers on his property and they in turn could keep chickens as a source of food. The cheetahs, Howard explained, had soon become wise to this fact and would rest overnight near one of the homesteads. Then at daybreak, they would each grab a chicken from the nearby settlement for an early morning snack. Looking at the dying embers of the fire, we were reminded that it was time to sleep. I still had so many questions for Howard but put them aside for the next day.

At daybreak I was woken and presented with a steaming hot cup of coffee. As I opened my eyes and reached out for the cup, I automatically looked towards the place where the cheetahs had been lying the previous evening. They had gone! Koba saw my blank expression and said, 'Don't worry, Ann. There they are, lying in that clearing, waiting for the sun's rays to warm them.' She indicated a sparsely bushed area close by.

I relaxed and turned to look to the east where the pale light of dawn silhouetted the large flat-topped thorn trees against the pink haze on the horizon. While I slowly sipped my coffee, a lone jackal wailed close by and the cry was answered by a call in the distance, barely heard above the loud chattering of birds. The crisp morning air soon cleared my eyes of all sleepiness and the splashing of cool, clear river water on my face revived me fully.

Suddenly we heard a sound that was almost bizarre in those wild surroundings – a cock crowing. Even though the bird was some distance away, the two cheetahs immediately sat up and I could read the expressions on their faces like a book: 'Ah, breakfast – what a good idea!' They were on their feet in a flash and began walking with a determined stride, obviously set on a particular destination. Then suddenly, without warning, they darted off and were swiftly out of sight. Howard had barely time to grab his wallet, realizing that he would have to pay for yet another expensive breakfast for the cheetahs. We all silently tumbled into the Land Rover, each one understanding only too well what was in store.

The nearby settlement was situated on the top of a rise and from the bottom of the hillock we helplessly watched the scene as it unfolded in the distance. The two cheetahs were stealthily nearing the huts, their long sleek forms clearly visible in the early morning light. Crouching low in the long grass they slowly approached. Women, oblivious of any nearby danger, were bent over a fire cooking a breakfast of mealie-meal porridge. Young children were playing hide-and-seek while waiting for their food. Father was a little distance away chopping wood. It was a typical rural scene. In the Land Rover we raced over the uneven veld towards the inhabitants, knowing that we could never reach them in time to prevent the drama that was about to unfold.

Then above the noise of the Land Rover's engine we heard the commotion: women were shouting in high-pitched tones, children were crying as they tripped and fell while clambering towards the safety of their huts. Suddenly, and only for a few brief seconds, all was still and then pandemonium broke out in the fowl run. Breathlessly we reached the huts, only to find an enraged chicken-owner gesticulating furiously and cursing loudly in language I fortunately did not understand. Occasionally breaking into English he repeatedly shouted, 'These two bloody dogs of yours are killing all my chickens!'

My eyes turned to two most unconcerned 'lads' who were under a tree nearby crunching the bones of a couple of hapless fowls. Howard handled the situation most diplomatically and quickly compensated the man financially for his losses. We felt that the situation was not conducive to an explanation as to how the cheetahs came to be wearing collars and had thus been labelled as 'dogs' by the poultry-owner. To the cheetahs, however, the chickens had merely been appetizers and later that day they brought down an impala ewe to satisfy their hunger.

Before returning to De Wildt, we discussed with Howard his future plans for the two remaining cats. Although Rodgers and Jan were scent-marking an area in the Klaserie Reserve, Howard felt that the test run had been completed and that we had nothing more to gain or learn from the venture. Accordingly, we now came to the conclusion that we could not release the two cheetahs permanently into the wild as we had intended: we could see that here, if they were not killed in retaliation for chicken thefts, they would be sitting ducks for any trophy-hunter in the area. And so it was decided that they should return to De Wildt. Howard enticed them with chickens into a small enclosure, and from there they were transferred to crates. He drove the 400 kilometres back to De Wildt non-stop and two hissing and spitting young lads greeted me at home when he arrived some six hours later.

Fortunately cheetahs adapt to this sort of change very easily and it was not long before the pair had settled back into the old way of life. I wondered whether they would remember their short taste of freedom and hanker after the unfettered lifestyle. A year later, in 1981, Rodgers was sent to a small game-farm in the Cape where he later fathered a number of cubs, while Jan remained at De Wildt with us, renamed 'Dead-Eye Dick' because of the loss of his eye. In many ways the experiment had been successful: we had undoubtedly learnt a tremendous amount from it but we realized that the biggest drawback was the fearless attitude that captive-born cheetahs have towards humans. Perhaps it would be possible, we mused, to isolate a pregnant female at De Wildt and once the cubs were born ensure that mother and young had as little human contact as possible. Food could be supplied by mechanical means and the group observed from a distance. In this way we hoped that the inborn fear of humans would remain with the cubs as they matured.

(37) A cub explores its environment.

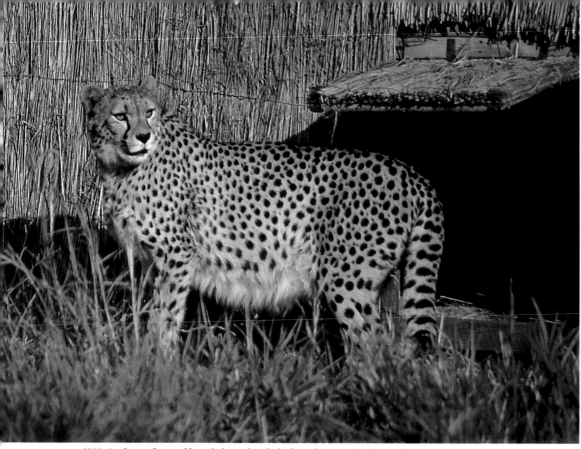

(38) Lady, in front of her shelter, shortly before she gave birth to a litter of eight cubs.

(39) Mr John and his mates create havoc with toilet paper and raw eggs.

(40) Mr John at the age of three years.

(41) The author with Mr John many years later.

(42) The coarse mane of a brown hyaena aglow in the setting sun.

(43) A young wild dog, bred at De Wildt.

(44) *After weeks of separation, litter-mates greet the female wild dog on her return.*

(45) *The leopard's pelt is marked with rosettes, very different from the cheetah's spots.*

(46) *The vulture restaurant attracts large numbers of meat-eating birds.*

(47) *A pair of blue cranes with chicks, foraging for insects.*

(48) Cape eagle owl.

(49) White-faced owl.

(50) Barn owl.

(51) Wood owl.

(52) *Cheetahs selected for release in the Pilanesberg National Park.*

(53) *Dead chickens were a regular part of the cheetah's diet.*

Howard summed up his findings by ending his report with the comment: 'This project certainly warrants further research. Additional experiments with captive-bred cheetah reared in the absence of human contact and the experimental release of a single adult female are needed in order to verify the findings of this study.'[†]

---

[*] H. L. Pettifer, 1981. 'Aspects on the ecology of cheetah (*Acinonyx jubatus*) on the Suikerbosrand Nature Reserve', in J. A. Chapman and D. Pursley, eds, *The Worldwide Furbearer Conference Proceedings*. Falls Church, Virginia: R. R. Donnelly, pp.1121-1142.

[†] H. L. Pettifer, 1981. 'The experimental release of captive-bred cheetah (*Acinonyx jubatus*) into the natural enviroment', in ibid., pp.1001-1024.

CHAPTER 12

A New Line of Thought

*A*fter Howard Pettifer's experiment, a nagging idea preyed on my mind: if only we could find an area where there was no human habitation and where our De Wildt-bred animals could live in natural surroundings; their offspring, if not interfered with, would then have an instinctive fear of people and this would make them less vulnerable to poachers.

It was early in 1980 that I heard talk of a new game reserve being established close to us in Bophuthatswana. An area of 60 000 hectares situated within an ancient and long-extinct volcanic crater was to be fenced and species once indigenous to the area were to be reintroduced. Of these species the cheetah was to be one. The walls of the vast crater formed a steep and rocky barrier and in the centre, beneath undulating grassy plains, the soil was very fertile. The area was dotted with typical bushveld shrubbery and water abounded in both natural lakes and man-made dams. Controlled and run by the Nature Conservation Department of Bophuthatswana, the Pilanesberg Nature Reserve, as it was to be known, was situated north-west of my farm a mere hour-and-a-half's drive by car. Adjoining this game reserve and next to a large artificial lake, a new luxury hotel and casino, known as Sun City, was being built by the hotel magnate, Sol Kerzner. In addition to many other attractions for tourists, trips would be organized into the Pilanesberg Nature Reserve. At first it was planned to site all accommodation and camping areas outside the reserve and only visitors in vehicles were to be permitted to enter it on day trips. This seemed to be an ideal situation for our cheetahs as contact with humans would be limited. Once more our hopes rose. On talking to the planners of the project, we found them to be very responsive when we presented our ideas. They could see no reason why our cheetahs could not be introduced into the new park and with the Pretoria Zoo's sanction to supply the animals, plans were soon set in motion.

Rhino, giraffe, eland, sable, kudu, gemsbok and impala were some of the species that had already been brought in by truck to stock the reserve. As these animals would have to settle and establish themselves before the big cats were released in their midst, an enclosure of approximately 10 hectares with a diamond-mesh fence and overhang was erected to hold the cheetahs. In the camp there was a koppie with an outcrop of great granite boulders. Large spreading indigenous trees grew among these rocks and provided summer shade, while the rest of the fenced area was grassland. We selected a group of 10 young cheetahs and delivered them in wooden crates a few hours before the official opening of the park. After the speeches, the doors of the crates were opened but the cats were reluctant to emerge from their darkened boxes. Once enticed out, however, they hastily made for the rocky koppie and soon disappeared from view.

We tackled this new project enthusiastically, helping and advising the game rangers whenever necessary. On our visits, we often found the cheetahs perched high on the koppie, looking wistful as they surveyed their surroundings, and I longed for the time when they could be released into the main area of the reserve. As food, a whole donkey carcass was provided every three days and this was always eaten rapidly. Slowly the cats adjusted to their new habitat. Later, in the spring after the first rains had fallen, the scorched earth changed into a green grassland and the trees were soon covered in new foliage. Pregnant ewes in the park were then close to dropping their young and summer was not far off. The change of season reminded me that the cheetahs had been held in the enclosure for almost a year, so I was thrilled to hear that a date had at last been set to release the youngsters into the reserve.

Overnight, however, our dreams were shattered. Plans for the reserve were suddenly changed and we were told that tented camps for visitors were to be built inside the park. When we thought of young children accompanying their parents on such a camping holiday we realized that we could not possibly take the risk of releasing our cheetahs into this area. The cats' fearless attitude towards humans was once again to be their downfall. There was, of course, the possibility that the cheetahs would move away from the areas of human habitation but we dared not take the chance. Reluctantly we decided to return the 10 to De Wildt and in their place to supply seven new wild-caught cheetahs that we had been holding in quarantine. These animals had been killing livestock on farms in Namibia and, rather than shoot them, the farmers of the area had donated them to the De Wildt project. We were sure that on release at Pilanesberg the seven Namibian cheetahs would soon take to the thick bush and thus be of no danger to visitors to the park. While advantageous in one sense, this plan also had a negative aspect: the chance that visitors would observe the seven cheetahs would be far less than would have been the case with captive-born animals.

After the change-over, reports of cheetah sightings occasionally came in from the Pilanesberg rangers and some time later, to our delight, cubs were seen. Soon, however, a note of concern crept in. We had hoped that the cheetahs would prey mainly on the prolific impalas, but unfortunately it was the ewes and young of other more valuable antelope that were disappearing. The half-eaten remains of sable and eland were found and we realized that the cheetahs had overstepped the mark and that economic considerations now had to dictate policy.

At the same time, knowing that the park could not survive on the income received from game-viewing alone, the authorities decided to set aside certain areas for hunting. As the reserve occupied fertile and arable land in a fast-developing country with a constantly increasing population, the management had to prove that the land allocated to it was being fully utilized and was commercially viable. Substantial revenue would be received from both trophy-hunters and the game meat obtained from the hunts. The authorities could not afford to lose rare and valuable antelope to the cheetahs, animals that were in any case seldom seen by visitors. In short, the cheetahs had become an embarrassment to the park.

This turn of events was a great disappointment to us. It was now apparent that suitable areas where this cat could be released were extremely scarce. I even wondered if cheetahs in the wild were doomed to extinction after all, not because of a tendency towards poor breeding but simply because of the lack of suitable habitat. Back at De Wildt, we realized that if we wished to continue the research work on the cheetah, our line of thought would have to change radically. Outlets would have to be found both locally and overseas for our surplus animals. After discussions at the zoo it was agreed to offer cheetahs to other recognized zoos and game-parks and we were greatly encouraged by the response we received. These institutions, by taking advantage of the knowledge and expertise that we had gained over the years, could now confidently establish their own breeding projects and in so doing increase the number of captive-born animals worldwide. Many wildlife species, such as Père David's deer, Derby eland and Arabian oryx, had been saved by international breeding programmes and at present many of these species are to be found only in zoos. This seemed to be the answer. As our cheetahs were born in semi-captivity, were accustomed to humans and fences and to being provided with food, they would adapt well to continued captive conditions.

In view of this change in the direction of our thinking, the staff of the zoo went ahead arranging documentation and air bookings, while the workshops busily set about making crates according to the specifications stipulated by airfreighters. Closed wooden boxes, approximately one-and-a-half metres in length by half a metre wide and a metre high, were made with well-ventilated sides. Provision for a container with water had to be made for long journeys, while a small inspection door was also a requirement. Zoos in San Diego, Cincinnati, Toledo, London, Amsterdam, Hanover, Stuttgart, and Melbourne were among the establishments requesting our animals.

As documents for each order were completed, the cheetahs were crated and dispatched, whenever possible on direct flights. Airlines did not require members of staff to accompany the felids on their journeys but they made the stipulation that the cats were to arrive at the air terminal three hours before departure. I was apprehensive, wondering how the animals would react not only to the long journey but also to the preliminary hours of waiting beforehand. However, any misgivings I might have had vanished when reports came back that the cheetahs had arrived safely at their destinations and that all were in excellent condition. I was interested to learn that wild animals, especially the felids, seldom need to be sedated and on long journeys are calmed simply by the darkness of their small crates.

Our attention now turned to the group of wild dogs. Three additional animals had been donated to us by the Transvaal Division of Nature and Environmental Conservation after a litter of pups had been rescued when their mother was shot in the Eastern Transvaal. With the existing four dogs and five bitches, we paired them off into enclosures, the unpaired female being placed in a camp with her sister. All went well and some months later in mid summer we noticed much activity and vocal interaction between the pairs of the group. With tails wagging and ears flattened, they uttered high-pitched whines or wails. They muzzled, licked and nibbled one another excitedly, showing their affection. The male often placed his head under the stomach of his mate, lifting her hindquarters off the ground; and whenever the female scent-marked certain areas in the camp, the male would immediately urinate over the spot. About a month later, mating was observed and we hurriedly erected dens adjoining, but outside, each camp. This structure, built of bricks, was a room approximately one-and-a-half metres square and two metres high, half of it being below ground. A narrow passage led into this darkened area and, with a drop-gate at the entrance and a movable roof, we knew it would be possible to check on the pups when the mother was outside in the camp.

After a gestation period of 72 days, one of the bitches gave birth to eight puppies. Mottled black and white in colour and weighing approximately 500 grams each, the youngsters were at first utterly dependent on their mother. Their eyes were closed but they were strong and healthy and clambered over one another to get to their mother's teats. A few days later, two of the other females gave birth. We heard the puppies whimpering whenever the mothers entered their dens and, not wanting to upset them in any way, we left them alone. There would be time enough to count and sex them later, I thought. Then, at feeding time on the morning after the birth, I noticed that the first litter was being guarded possessively by the mother – even her own mate was not allowed near – while the other two bitches showed no interest in their dens which were strangely silent. On investigating the two quiet dens I found nothing but one cold and lifeless puppy: obviously the females had eaten the others. I picked up the tiny animal and began to question what had happened. Research studies undertaken in the wild have found that only the alpha, or dominant, pair of a pack of wild dogs will mate and produce young, but I have also heard of two females in a wild pack that, together, mothered and fed the youngsters. Could it be possible that sub-dominant females intentionally kill their puppies soon after birth, and having milk, then help to feed the dominant bitches' offspring, thus producing strong and healthy whelps? I wished that we had a large fenced area into which we could release a pack of wild dogs and study them in more natural conditions.

There was a slight movement from the wet bundle I held in my hands and, realizing that the puppy was still alive, I rushed back to the hospital. The under-floor heating in one of the wards was kept on permanently for just such an occasion and in this warm room I placed the miserable youngster. I rubbed her vigorously until she was dry and then made up the usual mixture of milk and egg which we gave to newborn animals. As her mouth was fairly large, I selected for her feeding bottle a teat bigger than the one used for the cheetah cubs, and slowly encouraged her to drink the liquid. At first she swallowed unwillingly but soon there was no holding her back. I smiled as I compared the arrival of this puppy with the birth of the first cheetah cubs. How unprepared and inexperienced Godfrey and I had been then, and how easy things were now by comparison! I wondered how Godfrey would have reacted to the advances we had made and imagined the encouragement that he would have given.

Soon the puppy was quiet, and with a full, rounded belly was sleeping snugly tucked up within the cheetah blankets in a cardboard box on the warm floor. I telephoned Dave to give him the latest report. He was sad to hear of the loss of the puppies and noted that we would need to separate and recamp each pair, placing them out of sight and sound of the others. Wild dogs are seasonal, producing puppies once each year during the dry winter months of May and June, so we knew that we had ample time in which to rearrange these groups. I then went on to tell Dave of the puppy I had rescued.

'Be careful not to overfeed it,' he advised. 'They're inclined to drink far too much.'

'Would it be possible to take her to the house and raise her as a domestic dog?' I asked.

'I think it could be done,' Dave replied, 'but you'll have a terrible time coping with the offensive smell. Wild dogs have special scent glands that exude a very persistent and pungent odour.'

'Perhaps I could bath her regularly?' I suggested.

'It won't help. I think you should keep her in the hospital for a while and then we can see,' he said, tactfully deferring a decision.

The following day, on entering the puppy's ward, I caught a whiff of the musty smell that the dog exuded, and at once I realized just how unpleasant it would be to have that odour in my home. 'Stinky', as the female pup was soon named, took her bottle without much encouragement, and with a little coaxing urinated and defecated regularly. How much simpler it was to hand-raise her, compared with a cheetah cub! Strong and sturdy, she grew quickly and at 10 days her eyes opened. Girlie, my domestic dog, once again accepted the role of mother and took care of the puppy's toilet, nudging and licking her. Stinky, for her part, took it for granted that Girlie was her mother. The two became the greatest of friends and as the pup grew and became more boisterous they would play a rough-and-tumble game for hours on the grass outside the hospital. To Stinky the tennis ball was a great discovery and no matter where it was hidden she would find it. Even at this early age she was shrewd and crafty. Crawling submissively on her belly, she would creep up on Girlie making a twittering sound all the while. The moment Girlie was distracted, Stinky would wrest the ball from her mouth and then run off, disappearing into the nearby bushes. Girlie would sit back, looking amazed, clearly not used to these cunning ways.

The three of us would often go on long walks together, with Stinky flushing out every guinea-fowl and francolin from the long grass on the way. On our return, with Girlie panting heavily, Stinky showed no sign of being tired. Her stamina was remarkable, but as she grew older this was coupled with ever-greater nervousness and restlessness. One day she wandered off on her own and as I knew my neighbours would not appreciate the appearance of a wild dog on their farms, especially one that would raid chicken pens, from then on I had to keep her enclosed. She hated this and pined a lot, uttering a low distressed whoop-like call when I was not with her. When I entered her enclosure she would hysterically run round and round in circles, yapping all the time. Even Girlie's presence did not quieten or calm her. I discussed this development with Dave and we decided to move her to an enclosure adjoining the other wild dog puppies, thinking that in this way she could slowly be introduced to the pack.

The eight puppies, only a few days older than Stinky, were by now weaned onto artificial puppy-food and had developed well. It was amazing to see the control the adults had over their young; each time I watched the father firmly putting one of his offspring in its place I found myself thinking that some humans might well take a few hints from these wild animals. Over a period of three weeks, we introduced the eight puppies into Stinky's camp, finally bringing in the adults. The introduction went without a hitch and, although she was submissive at first, Stinky appeared to enjoy her new companions who played with her and accepted her as one of the group. She never forgot me, however, and I always received an unusually warm welcome whenever I went to the camp.

One morning three months later, while on the feeding round, I was shocked to find Stinky lying dead in the camp. She had obviously been attacked and badly bitten by the pack – for what reason I will never know. I picked up her body and sadly thought back on the many happy hours we had spent together. What a strange and endearing, but disorientated friend she had been. I realized once again how unfair it was to tame a wild animal, but at the same time I comforted myself with the thought that in this case I had had no option.

I often wondered what the possibilities were of releasing captive-born wild dogs back into their natural environment. I surmised that, as a result of the training and teamwork they needed to learn as youngsters, it was probably impossible to release adult dogs – but what about their pups? I have subsequently established through field studies that free-roaming hunting dogs will accept and adopt orphaned puppies. Perhaps, I thought, we could enclose a litter of our weaned youngsters in an area where we knew wild hunting dogs existed. I felt sure that the wild dogs would be attracted to the pups and possibly in due course would regurgitate food through the fence to them. When this happened we would know that the pups had been accepted and could be absorbed into the wild pack. We would, however, need an area where wild dogs were indigenous and where the authorities were sympathetic.

We discussed the idea with officials of the Natal Parks Board who showed willingness to attempt the release. However, they decided to include both adults and young animals in the pack which they intended to release in the Hluhluwe Game Reserve. The zoo donated a pair of four-year-olds and eight puppies and these were put together with another six pups received from a game-farmer in the Eastern Transvaal. After the initial three weeks, during which the dogs were held in a small quarantine enclosure, the pack was set free.

In the beginning a few losses were reported but now, 10 years later, a healthy pack of wild dogs is frequently observed in the reserve. It gives me a great thrill to think that some of the original animals of the pack were captive-born at De Wildt. Since then four of our puppies have been combined with adult dogs in Namibia and released into the Etosha National Park where one hopes they will also adapt and establish themselves.

Since the beginning of the De Wildt project, we had allowed very few guests to visit the centre as we felt that, with such a sensitive and at times purely experimental programme, peace and privacy for the animals were of prime importance. However, on a number of occasions it had been suggested that we should allow groups of schoolchildren into a section of the farm that was not in the breeding area. This was seen as an excellent opportunity to promote and teach some of the aspects of wildlife and conservation to the younger generation. We accepted the suggestion and decided that we could take children on a walking tour of De Wildt to view, in their natural surroundings, our wide variety of animals, many of which were rarely seen in the wild. We thought that the indigenous trees along the route could be named and numbered with small labels, according to family and species, and overall we would attempt to convey the importance of conserving flora and fauna and making youngsters aware of an irreplaceable wildlife heritage.

We also had plans to build an auditorium in which the guide, before setting out on the trail, could give the children a short talk on the biology and behaviour-patterns of the various animals that they would see. Information boards, diagrams and posters would be made available to assist the guide. But we needed a sponsor, and once again Dick Reucassel came to our assistance: he introduced us to Bill Bailey, Managing Director of Girder Engineering Company. Bill and his family are great supporters of wildlife and when we explained our plans to him, he offered to donate the building material and erect the structure for us. We were thrilled and grateful and slowly our ideas of conducting educational tours became a reality.

The local education authorities welcomed our proposal and circulars describing our plans were sent to schools throughout the Transvaal. We were amazed at the positive response we received; a date was soon set for our first group of scholars and there was a rush to complete arrangements. A path over the stony koppie had to be cleared, extra security fences had to be erected where the children were to stop to view the animals; trees had to be named and numbered and amenities such as toilets and a tuckshop to supply water and cooldrinks to thirsty visitors had to be built. Again, our good friends Koba and Hennie gave up their weekends and came to our aid. Dick spent many hours taking photographs and providing material to be used for posters and slides which could be bought by the children as mementoes of their visit to De Wildt. It was at this time that Eugene Marais joined us on a full-time basis. He was to lecture to the youngsters and accompany them on the walking tours.

The first busload of excited primary-school children arrived early one summer morning carrying cameras, notebooks and pencils, and set off on the trail with Eugene and their teachers. It was intensely rewarding to see the expressions on the faces of the children as they viewed a cheetah at close quarters – in most cases, for the very first time. When talking to and questioning the children from the large cities, I found it frightening to realize how little they knew of wildlife. Often we found that in their eyes the cheetah and leopard were one and the same animal. Wild dogs were invariably called 'hyaenas' and vultures were described as 'horrible birds' because they were thought to eat only rotten meat. Could we, in the few hours that were made available to us, ever hope to teach these children to appreciate the animals they saw on our walking tours? Within weeks my question was answered. I started to receive endearing and enthusiastic letters from pupils thanking us for giving them our time and for providing such an interesting and educational tour.

These outings have become so popular that now, over the weekends, we organize conducted tours for adults. Again accompanied by a guide, visitors are driven through the centre in an open vehicle and stops are made on the way to feed, photograph and discuss the various animals.

CHAPTER 13

Too Good to be True

*J*ust before the start of our cheetah-breeding season in 1981, Frank approached me. 'Ann, please don't breed cheetahs this year,' he said firmly. 'We're having great difficulty in obtaining fresh meat, and the demand for cheetahs from overseas zoos has dropped considerably. I suggest we skip this season and reconsider the situation next year.'

I naturally accepted his decision, but felt sad that we found ourselves in such an ironic situation: a mere 10 years ago, breeding cheetahs had seemed almost impossible and now we had to stop because of an oversupply. However, I was content with our considerable successes so far and especially with the thought that we had learnt so much about this secretive and complex feline.

My attention turned to two of Lady's daughters, Jean and Jumper, animals troublesome from the time that they had been weaned. To them a fence was erected solely to be climbed and I was never quite sure where I would find them when I went out each day on the feeding round. Fortunately, the perimeter fence was extra high and had a good overhang, thus making it impossible for them to venture onto neighbours' properties. While the two were young I tolerated their pranks, hoping that they would in due course outgrow their bad habit – but this turned out to be wishful thinking.

In the large enclosure across the stream, the herd of impalas had increased steadily in size and although I knew that I would soon have to reduce its numbers,

I stalled on the operation. However, Jean and Jumper, now adults and extremely adept at climbing fences, discovered the antelope on one of their unauthorized wanderings and decided to do the culling for me. As they climbed back into their own enclosure every evening, I was at first unaware of what was happening and only after finding the remains of an impala kill one day, did I understand what they were up to. The following morning I intentionally delayed the feeding round and, as I had expected, by the time I reached their camp the pair was missing. Immediately a search-party was formed and the two culprits, with full bellies, were duly found some distance away in the impala camp, lying in the shade of a large wild fig tree.

'You little blighters!' I exclaimed, laughing at the same time at their seeming expressions of scorn as, unmoving and not blinking an eyelid, they looked down their tear-lines at me.

Slowly, with the aid of the remains of their recent kill, I managed to entice the escapists into a portable enclosure and from there they were brought close to the hospital to a camp which had a three-metre-high fence.

The previous year I had noticed that the two sisters had taken a fancy to a male named Frik. He, however, was not particularly interested in youngsters, preferring more mature ladies; but this pair was not to be thwarted either by us or Frik, or by the high fence. One evening, after deciding it was time to look for a male, Jean scaled the barrier and romped down to the Monastery obviously in search of Frik. Ignorant of all this, I was disturbed to discover that Jean was missing from her enclosure when I checked the camps close to the hospital before feeding the animals the following morning. Thinking that she had decided once more on a change of menu, I called another search-party together and we spent hours looking for her in the impala area. It was all to no avail and, adding to our frustrations, the bush was thick and the grass tall after the recent rains. All the while I felt she was probably lying concealed somewhere nearby in the shade, stretched out after a good meal. The sun beat down on us unmercifully and I eventually called off the search, deciding to return later in the cool of the afternoon. I always tried to complete my feeding rounds in the early morning but by now it was unbearably hot and I quietly cursed Jean for the delay. I started to feed the others and, on approaching the Monastery, I could not believe my eyes. I burst out laughing when I discovered Jean there with Frik.

'You devil!' I said out loud. 'Is that what you think of our plans to stop breeding this year? Well, there's no point in taking you back to your camp now.'

I decided to let her spend the rest of her honeymoon in peace. Jumper was now on her own and a day later I was not at all surprised to find her, too, with Frik. I smiled when I saw His Lordship, looking very smug and superior, lying under a tree with his two admirers beside him. This escapade would probably mean two unwanted litters and I wondered how I would, in time, break the news to Frank.

A few weeks later I was approached by the owner of the Seaview Park, situated near Port Elizabeth, inquiring about the availability of cheetah cubs. This was just what I needed to solve my dilemma. I thought that if the zoo agreed we could supply one of the pregnant sisters to this game-park. The owner of the sanctuary was known to us and, having previously provided him with a tame cheetah, we were familiar with the good conditions in which his animals were kept. Describing the predicament I was in, I asked him whether he would be willing to take pregnant Jumper, but made it quite clear that she was a most proficient climber. He readily agreed to accept her and after the zoo had sanctioned the sale, Jumper was crated and swiftly dispatched on a direct flight to Port Elizabeth. Although I knew the departure of this troublesome cat would give me fewer grey hairs, I was still sad to see her go. She had a will of her own and I admired her for it. Once at the park, Jumper, still in early pregnancy, adapted to her new home without any further fuss.

Meanwhile, at De Wildt a good overhang was being erected on all sides of the maternity camp which would house Jean and where she was to be left alone to await the birth of her cubs. As her time drew near, her protein intake was increased and milk was added to her diet. Below her rib cage, her stomach was round and full, swaying slightly as she walked; her milk-line, too, was obvious. I looked at her, large and swollen, as she lazed in the sun awaiting her brood and I thought how difficult it must be for a free-roaming cheetah in this condition to run down prey. On the 92nd day of her pregnancy she refused food – which was for us a good sign – and started pacing up and down alongside the fence. The following morning she was in her grass hut and partially concealed from view. Although I could not see her cubs, I could tell by the relaxed expression on her face that she had given birth.

We left her in isolation, knowing that she would not attempt to leave her litter in search of food for at least 24 hours after the birth. I was secretly pleased that Jean had gone wandering and fallen pregnant as it was good to have cubs at De Wildt again. I knew that Jean, like her mother Lady, was a good mother and so I was not expecting her to abandon her offspring. On the next day's feeding round, Jean was waiting at the fence for her breakfast. Having asked Eugene to keep an eye on her while she ate her food under a nearby tree, I let down the flap to the front entrance of her hut and peeped in through the back window. This was a routine matter with newborn cubs as we always checked on their condition and made sure that they were warm and well fed. It normally took only a few minutes and the mother was none the wiser as she was then usually tucking into her food. Peering in, I saw five small furry balls, all close together and with distended tummies, fast asleep. But something arrested my glance: one cub, it seemed, was much darker in colour and marked differently from all the others.

'Watch it, Ann!' Eugene called suddenly. 'Jean's looking your way.'

I hastily slipped out, not wanting to upset her, my thoughts preoccupied with what I had just seen. It could not possibly be . . . we could not possibly be so fortunate. Eugene caught the puzzled expression on my face and asked me what was worrying me.

'I'm not sure, Eugene,' I replied slowly but excitedly, 'but I think Jean has given birth to a king cheetah.'

Jean had by now finished her meal, so that morning there was no further opportunity for us to observe her cubs. I quickly informed Dave who promised to call in on his way home. Dave and Miriam were now living on their farm some 10 kilometres west of my property.

Later that afternoon, when Jean had left her youngsters to go for a stroll, Dave was able to look in on her litter.

'I agree with you, Ann,' he said with a smile. 'There's certainly something different about one of the cubs.'

'I must telephone the zoo and tell them the good news,' I responded excitedly.

'No, not yet,' was Dave's firm reaction. 'Let's wait a couple of weeks and make quite sure.' Then he laughed. 'It'll be very embarrassing if the dark patches turn into normal cheetah spots.'

'You're right,' I rejoined reluctantly, wondering how I could possibly keep silent for so long. Deep down, I knew that we had a king cheetah and that the coat markings would not change.

Acinonyx rex, or the 'king cheetah' as it was then more commonly known, was almost a legend. It was said to have black stripes running down the length of its back with spots merging into lozenge-shaped blotches over the rest of its body. Apart from its coat colour, in every other aspect the king physically resembles the normal cheetah. The occurrence of this aberrant form of the common cheetah had first been revealed in 1926 when a Major A.L. Cooper had come across an unusual-looking skin in Rhodesia (now Zimbabwe). He had thought that the pelt was that of an unknown felid and was convinced that he had discovered a new species of cheetah. He had sent a photograph of the skin to Mr R.I. Pocock, then the curator of the Mammal Section of the British Museum in London, and had asked for his opinion. Pocock had at first thought the animal could have been a mutant or variant form of the leopard, but later, on receiving an actual skin from Cooper, decided that it was a new species of cheetah and named it '*Acinonyx rex*'. From 1926 to 1975 there had been six recorded sightings of such an animal, while skins brought in numbered twelve.* Seen only in the triangular-shaped area made up of the Eastern Transvaal, Zimbabwe and Botswana, the last photograph of a king cheetah had been taken in 1974 in the northern section of the Kruger National Park. Many theories existed about the strange colour variation and it was widely thought that this cheetah was a separate but disappearing species. Another opinion was that the animal could be a cross between a cheetah and a leopard.

In the 1970s an English couple, Paul and Lena Bottriell, had come to southern Africa on an expedition aimed at solving the mystery. After spending two years on their search and with nothing forthcoming, it seemed the king cheetah would remain one of nature's secrets. But here, at De Wildt, Jean in one naughty escapade with Frik seemed to have proved beyond doubt that the king cheetah is indeed an abnormally marked variant of *Acinonyx jubatus*. In other words, it was not a separate species but a cheetah with a pelt that had a colour variation, the phenomenon being made possible by a single recessive gene carried by both parents. There were no special or unusual markings on Frik's or Jean's coats to suggest that they carried this particular gene, and to Jean the strangely marked cub was no different from the others in the litter. 'The King', as we called him, was from the start brought up quite normally with his brothers and sisters and received no special treatment.

Suddenly I thought of the other pregnant female, Jumper, that had been sent to Port Elizabeth. She was Jean's sister and had mated with the same male – could she also give birth to a king? Immediately I decided to call the Seaview Park but, not wanting to raise the owner's hopes, I did not mention Jean's achievement and casually asked if Jumper had given birth.

'No,' came the reply, 'but she refused her food this morning, so I expect her to go into labour tonight. Is she likely to have a problem?'

'No,' I answered. 'She's had cubs before. But do please let me know how many youngsters are born.'

Early the following day I answered the telephone to hear an excited voice saying, 'Ann, I just had to tell you that Jumper's given birth to three cubs. But, it's odd – one of them has strange markings. Jumper's lair is some distance from the fence so it's difficult to see, but one cub is very dark. From what I can make out, it has broad dark stripes down its back. To me it looks like a king cheetah if I compare it with pictures I've seen. But surely this isn't possible? What do you think?'

'We have a similar situation here,' I replied quietly, trying to suppress my excitement. 'One of Jean's litter of five is also very dark and could possibly be a king cub. But Dave has suggested that we remain silent until we are absolutely sure of the fact.'

'That'll be difficult,' said Jumper's owner, laughing. 'But I'll try. Dave's right, of course.'

As I replaced the receiver a thought crossed my mind – I had learnt yet another lesson: never sell a pregnant animal!

That we were fortunate enough to breed this rare and magnificently marked animal was an unbelievable and unaccountable stroke of luck – or was it? Perhaps something in nature had preordained this new development? Whatever the reason, we had now been given the key to the breeding of the elusive king cheetah and it was up to us to take full advantage of the fact.

And what of the future? With a nucleus of king cheetahs established at De Wildt, could an area of land possibly be set aside where only king cheetahs would live?

As it was necessary for both parents to carry the king cheetah gene, we now went through all the records and traced the one carrier through Jean to Lady, on the female side, while Frik, of course, we knew to be the male carrier. But Lady and Frik had been with us for some time, so why had the mutation not appeared earlier? The answer was simple: sexually Frik was a very lethargic animal, who had always been slow and uninterested in the ladies. As there were other much more virile and active males, Frik had been pushed aside and had seldom been selected for breeding. Fortunately, though, he had fancied Jill, and from these matings we had had cubs to retain the bloodlines of litters which had been born before Jean and Jumper's escapade. But Frik was now 13 years old, already in a camp with the other older cheetahs – 'pensioners' we called them – and would not be with us much longer so our efforts to breed king cheetahs would have to turn to his offspring.

We examined Lady's, Frik's and Jean's coats individually, looking for rows of spots on their backs or larger-than-normal markings on the rest of their bodies but there were no tell-tale signs that showed physically that they might carry the necessary king cheetah gene. We concluded that only through breeding and the arrival of new offspring could we be sure that the parents were definite carriers.

A couple of years later, long after the excitement of the first king cheetahs had died down, two of Frik's sons were brought into the breeding programme when they reached maturity. One mated with Jean and she produced five cubs, two of which were kings, but unfortunately all were dead at birth. The other son mated with Cynthia, a younger daughter of Lady, and she produced four healthy cubs – one of which was a king. I was thrilled with this confirmation that these parents were gene carriers and that they could be added to our breeding group. At the same time I knew that we would have to be careful to avoid inbreeding, but once we had established a nucleus of kings and had identified the gene carriers we would be able to start mating these animals with completely unrelated cheetahs and in so doing introduce new blood. As this was Cynthia's first litter, I was unsure of her maternal instincts and, because of the possibility that she might abandon her litter, I decided to hand-raise the valuable offspring.

During this period, at the game-park in Port Elizabeth breeding prospects were not as good as at De Wildt. Jumper and one of her offspring had died of anaemia and as the owner intended selling the park, the Pretoria Zoo offered to take over the female king – 'Queenie' as she had been named – and her brother. The offer was accepted and in due course the zoo handed over to us the two cubs.

While all our efforts were concentrated on establishing a good breeding group of king cheetahs, suddenly two young king cubs were seen, with their normal-coated mother, in the central area of the Kruger National Park. Soon after this, a lone 'queen' was sighted not far from the first group, on Hans Hoheisen's farm which adjoins the park. Since it was on this farm that our trial release of the three male cheetahs had taken place eight years previously, I wondered whether there could be any connection. Perhaps, between these newly found kings and the three sons of Lady that we had attempted to release, there may have been links. The possibility was there but we would never be sure. I was nevertheless very pleased that king cheetahs had been sighted in the wild, as it proved that their coat patterns were a perfectly natural occurrence.

Then, unexpectedly, Marie du Plessis, the wife of a game-farmer in the Eastern Transvaal contacted me. 'Ann, I have a six-month-old cheetah cub for you,' she said. 'It's a young male king that was found in a poor condition in the veld and was given to me to hand-raise. Now I feel it's time to send it to a permanent home. Would you be prepared to take it over?'

I was dumb-struck. A king – what a wonderful offer, for it would bring new blood to our existing breeding stock.

'Are you sure it's a king?' I asked. 'King cheetahs are so rare. It's not possibly a cheetah with a very dark coat?' I probed hesitantly.

Marie laughed. 'No, it's definitely a king,' she said with conviction. 'And I've been offered a substantial sum of money for it. But I would prefer the youngster to be given to you.'

Still amazed at this stroke of luck, I replied, 'If it is a king it'll be a tremendous asset to us in our breeding programme. I'll never be able to repay you for your kind offer.'

But Marie said that as long as the cub had a good home, she would be happy and I assured her that I would contact the zoo and arrange permits at once.

Still with a slight flicker of doubt in my mind that it could be a normal cheetah with just a very dark but otherwise normal coat, I told Dave the news. He was as pleased as I was, and instructed me to arrange immediately that the cub be given the feline enteritis vaccine and advised me to place the newcomer in quarantine – away from the other cheetahs – for three weeks, until the vaccination had taken effect. As soon as the zoo had arranged the necessary documentation, I sent Eugene off in the Kombi on the 400-kilometre journey to the game-farm, armed with the vaccine and the crate necessary to transport the cub.

Impatiently I remained behind, occupying my time with normal daily duties and eagerly awaiting Eugene's return. The following day he was back and greeting me excitedly he exclaimed, 'Ann, it really is a king!'

Then I watched with delight as a magnificent king cheetah cub, in perfect condition, stretched himself in leisurely fashion as he emerged from the crate.

'He's very tame, and because of the dark coloration of his coat he's been named "Spookie",' Eugene said. 'And I have his blanket, ball and food-bowl with me!'

My immediate response was to telephone Marie and to thank her once again for her gift and to congratulate her on the superb condition of the cub – and at the same time to apologize for my initial doubts as to the animal's 'pedigree'.

All the rooms in the hospital were then occupied by civets, servals and caracals, so I decided to allow Spookie to share my home with me for the quarantine period. A small portable camp was quickly erected outside to enclose him during the day, and, as it was only cat-contact that we were avoiding, in the evening he was allowed to romp and frolic with the dogs in the house. Scruffy, the maltese poodle, was Spookie's favourite. Round and round the furniture they would chase each other and whenever Spookie was losing the race or was tired, he would jump onto the piano where Scruffy could not reach him. Tug-of-war was another popular pastime and then I had to make quite sure that my clothing and shoes were well out of reach. But I enjoyed having a tame cub with me again, and all bad deeds and mischief were forgotten when Spookie came to me purring loudly. As I was never sure of what pranks he would get up to during the night, I locked him in my study, after having first covered the entire carpet with newspapers. Then I placed his blanket on the top of my desk as its height gave him a sense of security. Although he was content playing with the dogs I realized that he needed cub mates and was pleased when the three weeks of quarantine were over and he was allowed to be introduced to cheetah youngsters of the same age.

Slowly over the years we have developed our 'royal blood' strain. As these are the only captive-born king cheetahs in the world, I feel that our breeding stock should be increased and truly established before the animals are exported to other institutions. The king has a long-sought-after gene pattern and it would be a shame if the bloodline were for some reason to disappear.

* See Wrogemann, *Cheetah Under the Sun*, pp.17-21.

CHAPTER 14

Chaos!

After a brief cold snap in the third week of September 1986, summer was truly upon us. The days went by with no rain in sight and the hot sun scorched the dry earth. New plant growth that had blossomed after the short early rains, soon withered and died and it seemed that another dry season lay ahead. Now into our fifth consecutive year of drought, I sometimes wondered whether the weather pattern was changing and if our part of the land was destined to become a desert. The spring on the farm had dried up and the level of the water table in the boreholes was dropping steadily – a bad omen. The months passed with no sign of rain until early December. Then the weather changed. Although it was still blisteringly hot during the mornings, a high wind and dark storm clouds gathered in the late afternoons. But still no rain fell – until 10 December, a day that I shall never forget.

The sun that rose that Thursday morning was a great red ball of fire in the east and the few small clouds that hovered in the sky soon disintegrated with the intense heat. It was easy to see that it was going to be another swelteringly hot day, with the temperature rising again into the high thirties. As the day wore on, many of the animals at De Wildt lay exhausted and listless, panting in the sparse shade of the trees. With little enthusiasm, we continued to attend to our daily duties and in the afternoon, at about four o'clock, Eugene and I started our usual round that involved feeding and locking up the cheetah cubs and wild dog puppies for the night. Ominously heavy clouds had by then begun to form in the west, soon obliterating the sun but, to our relief, cooling the air. We looked up at the sky and silently prayed that this time it would rain. By five o'clock the dark bank of clouds looked evil and threatening, taking on a sickly green colour as it rolled towards us from the south-west. Then a gusting gale-force wind started, churning and lifting the dry soil from the ploughed fields into a stifling duststorm that swept visibly along the valley. Never before had I observed such a strange sky; I quickly ran to the hospital to drive the last lot of cubs indoors. As always, the youngsters thought I was playing a game and mischievously ran backwards and forwards, refusing to be locked up. I cursed them under my breath but eventually managed to get them inside. As the last one was carried indoors, I heard the loud clattering sound of hailstones as they fell on the corrugated-iron roof of the hospital.

Then, suddenly, all was quiet again. I ran from the hospital to the house and had just reached the front door when it seemed that all hell had broken loose and a solid wall of hailstones the size of golf balls came hurtling down. Helpless and aghast, I listened to the sounds of window panes around me being smashed and the hail, driven by the strong wind, pelting into the rooms. After bouncing and sliding down the thatched roof, the ice started to accumulate around the base of the house, forming a barrier. Soon the level of melting hail, together with rain-water, rose and seeped in under the doors. At the time I hardly noticed this as my thoughts were with my animals. Would they, could they, stay alive through such violence?

It seemed an eternity passed, during which the deafening roar continued and the storm progressed on its path of chaos and devastation. After 20 minutes – which felt like hours – the volume of hail lessened and the rain took over. The ravished earth, now devoid of all its grass covering, could not hold back the deluge of water which gained force as it raced down the mountainside, carrying loose stones and debris with it. I tried to venture outdoors but it proved impossible in the heavy rain and amidst the hail, which by now lay half a metre thick on the ground. The telephone line was dead, so I could not call for help. I knew, in any case, that next door Eugene would be in just the same predicament and complete-ly housebound. Eventually the rain eased and although it was still teeming I crunched my way through ice and debris, to the young cheetahs enclosed in the lower camps.

I was shocked as I worked my way there. Not a leaf remained on the trees and not a blade of grass was visible. Although warmly clad, I shivered as the cold air penetrated my thick winter jerseys and my feet were numbed by the icy water that seeped through my shoes. The cheetahs were huddled together, shivering and trying to keep one another warm. At least they were still alive, I thought with relief. But the young king cheetah which I had hand-raised was not as fortunate. I found him bogged down in the ice, unable to move, for he had clearly panicked and had tried to escape, only to tire himself out. He was now exhausted and almost unconscious and so I carried him quickly to the hospital and into a ward where the under-floor heating was on. After being rubbed down rapidly and briskly to increase the speed of his blood circulation, he slowly started to purr and then to lick my hand as if in gratitude. I knew then that I was over that hurdle, but by now the last light of day was fading and I had no time to lose in attending to all the other animals.

Soon Eugene arrived, having first made temporary arrangements to contain some of the damage in his house, and together we started out with torches to check on the other animals. After searching the adult cheetah camps we gave up in despair: the enclosures were large and the few cheetahs we did see ran from us in panic. We then set off in the truck to see to the wild dogs and the male cheetahs in the Monastery at the back of the koppie. On the way we passed the owl cages and saw that the nylon netting that enclosed them was torn to shreds and that a number of the owls were caught in the tattered mesh. We had fortuitously brought cardboard boxes with us and slowly we collected the dazed birds. The two little pearl-spotted owls were dead, but the others, despite the shock and cold, were alive – it was obvious that the netting had protected them to a certain degree.

We hurriedly returned the boxes to the hospital and resumed our journey. As we reached the road that ran along the dam wall, Eugene slammed on the truck's brakes – he had suddenly become aware that very little of the wall remained. It turned out that the overflow pipe had become blocked with debris and the water behind it had risen in the dam, eventually flowing over the top and eroding the soil of the earth-wall. We tried to make a detour by using a neighbour's private road but we were again thwarted as this route also was impassable. By now it was well after one o'clock and we realized there was really nothing constructive that we could do until we had the light of day to help us. Deciding to snatch a few hours of sleep, if this were possible, the two of us arranged to meet at sunrise.

Despite my exhaustion, I slept fitfully that night and with the first light of morning I sensed, with a cold shiver, that something was sorely amiss – but I could not fathom what it was. I dressed quickly and went outside wondering what was so strange. Then I realized – it was all so quiet. A deathly silence lay over the scene of devastation and I identified that what I was missing was the sound of birds with their early morning song. Not one of our wild garden birds remained. There were no noisy bulbuls, no busy weavers – not even the lively sparrows and sedate doves were to be seen or heard. But on looking down at the ground, I found the birds, one by one. They lay flattened at my feet among the battered leaves and debris. Hail still lay thickly on the ground and ghostly white trees, denuded of all their leaves and bark, stood starkly against the soft morning light.

I stood speechless. I could neither scream nor cry. All that had been built up over the years was smashed and I had no emotion that could meet this disaster. Trees and bushes that I had cherished were now barren sticks pointing to the sky. 'They will recover. They *must* recover,' I avowed fiercely as I made my way down to the hospital. The young king that I had revived the evening before welcomed me with enthusiastic purrs, and while giving him a quick hug I mentally reminded myself to ask Dave about administering antibiotics. Cheetahs are very prone to pneumonia and after the sudden and extreme cold that we had had I knew that anything was possible. Just then Shironga arrived to say that a long stretch of the fence at the back of the koppie was down and that a number of the impalas were on the neighbour's ground. I questioned him further, but he could not say for sure whether cheetahs, too, had escaped. While I was talking to Shironga, Dave arrived. Although his home was only 10 kilometres west of De Wildt, fortunately that area had not been in the path of the hailstorm. However, on passing our gate as he did every day on his way to and from work, Dave had been shocked by the scene of devastation. He could not stay to help us but he promised to alert the zoo as soon as he got to work, and to arrange for Dr Richard Burroughs, one of the zoo veterinarians, and other members of the staff to come to our assistance. Dave promised to return as soon as he could.

While Eugene checked on the animals in the female breeding camps, I went with Shironga to assess the damage done to the fence. From the top of the koppie the path of the storm could be clearly seen, a band approximately three kilometres in width that had come over the north-facing mountain behind us and had smashed its way in a plainly visible strip across the valley below.

The neighbours must be in just as serious a plight, I thought. Margaret Roberts, the well-known herb specialist, lived on the adjoining farm. She had taken years to build up her flourishing herb garden which I surmised must now be completely destroyed. On arrival at the fence, we were faced with a mind-boggling scene of storm havoc. Dead adult impalas and their newly born lambs lay everywhere amongst the flotsam and uprooted plants. Between the carcasses were strewn dead guinea-fowls, hares, birds and reptiles. The hail had packed up against the perimeter fence, breaking the binding wire from the poles and now Shironga and I were confronted by long stretches of netting which lay on the ground. The diamond-mesh fencing spanning the stream was a tangled mess of wire and twisted poles, and tree stumps that had once required at least six workers to lift them had drifted down in the floodwater, smashing all barriers in their path and ending in the dam below. But there was no time to ponder on the extent of the damage – what we needed was help and we needed it quickly.

On our return to the hospital we met Eugene with a load of carcasses – a wet mass consisting of six cheetahs, two caracals, two crested cranes, three blue crane chicks, a nyala and a flock of guinea-fowl lay in a heap on the back of the truck. My reaction was single-minded: the fences had to be repaired speedily and we had to ascertain whether any of the surviving animals were injured. Within a couple of hours Richard Burroughs and members of the zoo staff arrived to give assistance. After first viewing the devastation, they decided that the priority was to erect temporary fences at strategic points as quickly as possible.

Leaving others to get on with this job, Richard and I went from camp to camp checking on the condition of the remaining animals. I knew that towards the top of the koppie there was an area containing a great deal of ironstone which during turbulent electric thunderstorms in the past had often attracted lightning. Here, during this storm, the lightning had struck a tall, spreading syringa tree, under which three male cheetahs had sought refuge. One of the three had escaped unharmed, while the second we found to be partially blinded and with deep burn-marks across its rump. The third had been killed outright. Slowly coaxing the blind and confused animal into a crush, Richard attended to its wounds and did his best to ease the pain it was obviously feeling in its eyes.

Further up the koppie we discovered that a tame female cheetah was missing and after calling for her and searching the enclosure she came up to us slowly, limping. Big cuts and gashes scarred her legs and there was blood on her mouth but as she was tame it was easy to crate and transport her down to the hospital where Richard could examine her more closely. There it was established that no bones were broken, so after attending to her lacerations she was placed in a ward in the hospital to recover from her ordeal. I often wondered afterwards what had really gone on in the enclosures during the storm – had the animals run about in frantic panic or had they lain flat on their bellies, under a tree or bush, trying to escape the hail? I will never know, but whatever their reaction, I am sure that it must have been a most terrifying experience for them all. Among the remaining cheetahs we saw no visible injuries – though bruises there certainly were. The wild dogs and hyaenas were unharmed as they had obviously been able to take refuge from the deluge, down in their dens, and the vultures, perched on an old spreading tree without covering, had miraculously weathered the battering.

As Richard and I drove to the back of the koppie where the workmen were hard at it repairing the fences, suddenly two impala ewes ran at great speed across the road in front of the vehicle. Just as I braked to avoid them, six cheetahs appeared through the bush in hot pursuit. I sat speechless with amazement until Richard's voice broke the silence: 'Ann, we've got a big problem on our hands. Where on earth have these cheetahs come from?'

'Those are the six males from the enclosure across the stream,' I replied. 'There must be an opening in the fence there that we overlooked,' I added dejectedly.

By now the six had brought down one of the impala ewes and were in the throes of killing it: one of the cheetahs had sunk its teeth into the animal's throat and was applying the typical strangulation hold. Realizing that the two of us would be unable to stop the slaughter we raced back for help, on the way calling Eugene over the two-way radio and asking him to round up as many of our workers as possible. On reaching the hospital, we found helpers already assembled and waiting and after everyone had hurriedly climbed aboard the truck we rushed back to the six escapists. It was not difficult to locate them as by then another impala had been brought down quite close to the previous kill and they were now eating off that carcass. We formed a long human chain and slowly herded the errant and unwilling cheetahs back towards their camp. From time to time, one or other would turn on its tracks, making a bee-line for freedom and there was no way that we could stop such a determined cheetah from breaking through our human barrier. Uttering audible curses, on each occasion we were forced to turn once more and clamber up the koppie following the runaway. Although their stomachs were full, they were not at all eager to return to their enclosure and only after hours of patience and perseverance on our part were we able to herd them all in. We soon found the broken section of the fence and quickly pulled up the flattened mesh. I smiled to myself: I had to admire those six big cats, for despite their years of confinement and after being hand-fed for so long, there was no doubt that they knew exactly what was under the sleek, smooth skin of the impalas.

With the knowledge that the surviving animals were as well as could be, we started clearing away the storm debris, replacing fences and restoring order. It took many months of continuous work before we could return to our normal duties. As the days went by I noticed that the general weather pattern of the past five years was changing, for good soaking rains now fell regularly and the barren land slowly returned to its natural state. As always, time heals and two years later the spring of 1988 saw the trees and bush that were once stripped of leaves and a great deal of bark, burst forth with lush thick foliage and flowers. The large feathery-leafed acacias displayed a mass of fragrant fluffy yellow balls; the sweet-scented ochnas adorned with small buttercup-like flowers were humming with nectar-seeking bees and insects, and the wild pear trees were covered with clusters of creamy white blossoms. New birdlife soon moved into the area and it was not long before I woke once again to its noisy song. But in the midst of this glorious display of new plant growth, brilliance and colour, bare and debarked dead trees still stand even today – some years later – as stark reminders of that awesome storm.

CHAPTER 15

New Projects

*I*t was sad for all at De Wildt when both my good friends Frank Brand and Hannes Koen were compelled to resign their positions at the Pretoria Zoo because of ill health. Frank retired first, in December 1984, and Hannes followed a few months later in May 1985. Over the years they had constantly been at my side to offer help and I shall always remember and value the friendship and support they gave me. The position of director at the zoo was filled by Willie Labuschagne who until that time had been curator of the zoological gardens in Johannesburg. With fresh ideas and goals, Willie tackled his task with great enthusiasm. At De Wildt, for example, two new projects were soon under way and a completely new field was opened up with the establishment of a breeding nucleus of suni antelope and the rare and very endangered mammal, the riverine rabbit.

The suni is a small and dainty antelope, an adult weighing between five and seven kilograms. The coat of the animal is a smooth dark reddish-brown colour, fading to white on the under-belly; the underside of the black tail is also light in colour. Only the males carry horns – short, heavily ridged and sloping slightly backwards. It is a delicate animal with thin, short legs. As it is shy and secretive, it is rarely seen in the wild and if detected or disturbed it will freeze its position for a few seconds before speeding off into the nearest thicket. Sunis are mainly browsers, feeding on leaves and young shoots in the early morning and late afternoon, and resting during the heat of the day. The animals are found solitarily, in pairs, or in family groups. Females give birth to a single lamb after a gestation period of approximately 220 days, and the male is known to mate with her soon after parturition.

Although in theory sunis range in distribution through Kenya, Tanzania, Malawi, Mozambique and northern Natal, in 1981 the Natal Parks Board was concerned about the gradual disappearance of the species from their game reserves. A researcher, David Lawson, sponsored by the Endangered Wildlife Trust, was appointed to investigate the state of the suni in Natal and to find possible reasons for its dramatic drop in numbers. After a four-year study, David published his findings* in which he put forward two significant causes for the antelope's population decline. First, there was the destruction and removal of the suni's natural habitat which had been cleared for farming activities; and secondly, there was the uncontrolled population growth of the nyala – a much larger antelope, but also a browser, that removed suni-height vegetation and opened up the dense cover, making the smaller animal easy prey to leopards, pythons and eagles. As sunis are difficult to study in their densely vegetated natural environment, David had held several in captivity for observation. On completion of his work these animals were offered to De Wildt as a breeding nucleus, and after discussions with Willie Labuschagne we accepted them willingly.

On David's advice, we erected three bomas which were quite different from our predator camps. The perimeter fences of these enclosures were made of wooden slats, running horizontally and nailed to strong wooden uprights and closing off a thickly wooded area 25 metres by 10 metres and two-and-a-half metres high. In one corner a night-room two-and-a-half metres square was built, and here the antelope were to be held for the first few days after their arrival and where, afterwards, they could rest or hide during the day. On completion of the camps, David made the journey to De Wildt by truck, travelling through the night with the sunis boxed singly in wooden crates, and arriving at the farm early the following morning. On his instructions we had cut branches off trees and bushes which grew at De Wildt but which also occurred in the Natal suni habitat – plants such as the buffalo thorn, *Ziziphus mucronata*, the raisin bush, *Grewia flava*, and the white stinkwood, *Celtis africana*. We placed these fresh cuttings in the night-rooms, together with lucerne and antelope cubes – the latter being a nutritious and well-balanced artificial food which is fed to herbivores in the zoo.

David brought us seven sunis – three males and four females – which were released into the three camps immediately upon their arrival. The small antelope had been accustomed to captive conditions and did not appear to be stressed by the journey for they settled in readily and accepted the food we offered them. To our great delight three of the females gave birth within the following six months, each to a single lamb. The remaining female, however, was still too young to reproduce. These newborn lambs were beautiful – so tiny and dainty – standing only approximately 25 centimetres high. The mothers hid the youngsters in the tall grass of their enclosures, returning two to three times a day to feed them. After only a week the lambs had already started nibbling on the dried leaves lying on the ground.

We were very excited by this progress but we all knew that the group of sunis was too small to establish a good breeding herd, so I suggested to Willie Labuschagne that the zoo should approach the KwaZulu authorities with a request that they catch eight more sunis for us in their Temba Elephant Reserve. This appeal was readily complied with and Eugene travelled to the reserve to assist with the capturing of the antelope. Within the following three days, four males and four females were caught and airfreighted to Jan Smuts Airport. On arrival, each consignment was taken directly from the aircraft to a waiting vehicle which left at once for De Wildt where the antelope were released into the bomas – no more than 12 hours after capture.

When caught wild, antelope experience tremendous stress and shock which, if too extreme, can cause sudden death. According to a report investigating this phenomenon, death 'occurs within one day to four weeks after capture, but usually within one to two weeks'.[†] The occurrence of stress of this kind is one of the main hazards in the capture, transportation and translocation of wild animals and it affects most species. The condition is known as 'capture myopathy', or 'white muscle stress syndrome'. With this in mind, we found the two weeks following the arrival of the sunis a very tense and worrying time and we did, in fact, lose two males during this period. After the first week, the newcomers started nibbling on the fresh leaves of the indigenous plants, and these we then renewed twice a day. Once it was found that they were all eating well, we slowly introduced lucerne and antelope cubes to their diet. Today, several years later, we have a healthy population of 35 animals and the zoo has already donated seven to the National Parks Board to add to their breeding nucleus at Skukuza in the Kruger National Park. It is intended to breed up a large population there and at a later stage to release a certain number yearly into the areas of the park where sunis were once plentiful.

Soon after the suni project was established, Professor John Skinner approached both me and the zoo and asked whether we would consider trying to breed the rare and elusive riverine rabbit. I was thrilled with the idea. Until recently very little was known about this small mammal but I did know that it was one of the most endangered species of the subcontinent. Found naturally in dense riverine scrub which grows along the dry river beds in the western Karoo of the Cape Province, the population of this rabbit had dropped drastically over the years, following the destruction of its natural habitat. Areas where the species occurred have been cleared for the cultivation of irrigated crops, and dams have been built to hold back water when rivers are in flood, thus destroying the riverine bush. In the late 1970s Professor Terry Robinson, a zoologist from the Mammal Research Institute in Pretoria, had investigated the status of the rabbit and his findings even then had been very disturbing. The number of rabbits was steadily dropping and in some areas the species had disappeared completely. Terry's project was continued further by Andrew Duthie who, realizing that the animal was nocturnal and difficult to follow, decided to fence off an area where he knew the rabbit lived. For four years he concentrated his studies on the riverine rabbit population in this large enclosure, observing its behavioural patterns.

Measuring about 43 centimetres in length, the rabbit's most striking feature is its exceptionally long ears which are fringed along the inner edges with short white hair and tipped with black. Its coat is soft and silky and greyish-brown in colour; a dark brown stripe on the head runs along the sides of the lower jaw broadening and disappearing near the base of the ear. The tail is round and bushy and the legs short with thickly furred feet that serve to give it a firm foothold on soft river sand. Extremely nervous by nature, the animal lies up during the day in a form, or slight depression, in the flattened grass, absolutely motionless and with ears pressed backwards. The colouring of its body blends in with that of the Karoo soil and it will flush or scurry out only if it senses danger. When nearing parturition, the female will start to dig into the earth and make an underground nest in the bank of a dry river course, lining her future breeding stop, or burrow, with fur plucked from her body. The opening to the burrow is then covered with loose grass which is temporarily removed by the mother when she enters the stop to allow her young to suckle. Known as 'kittens', her young – usually two in number – are born utterly helpless and naked. They are quite unlike the common hare's young which at birth are fully haired and, with eyes already open, are able to move about soon afterwards.

A large enclosure – 100 by 50 metres – was erected for the rabbits at De Wildt. It was situated on a level grassy area and was fenced with bird-wire to keep out small predators; in addition, it was covered with plastic netting to protect the young rabbits from birds of prey. The grass in the camp was tall and thick, providing good protection for the animals, and alongside it we had planted beds of lucerne on which they could nibble. When the paddock was ready, Andrew arrived with four bucks, or males. He wanted to observe their reactions to the change of habitat before introducing the does, or females. On release, the four were left in isolation but, to our great disappointment, the following day two were found dead: later a post-mortem examination derdermined that they had died of exhaustion. Once again we were reminded that newcomers ought first to be held in a restricted area and given time to adapt and calm down. A small portable night-room was quickly built and into this the remaining three rabbits – one buck and two does – were placed. Apart from a short visit from one of us twice daily to renew the fresh lucerne, the enclosure was placed strictly out of bounds. Then, three days later, the hatch of the night-room was left open to allow the rabbits to come and go as they pleased. This was successful and after that, apart from the walk-through that Andrew, Eugene or I did once a week to check on the condition of the rabbits, the camp was strictly isolated.

A year passed but no kittens were sighted. Was this an animal that would not breed in captivity? Why, when domestic rabbits were such prolific breeders, was this relative so different? Early one morning in September 1988 I accompanied Andrew on his walk through the enclosure. We spoke very little, and I sensed that Andrew, who had always been optimistic in the past, was slowly losing heart. Suddenly he froze. 'Ann,' he called softly. 'Move slowly over this way, towards me.'

I did as he instructed and there hidden in the lush green lucerne below me, were two small furry kittens, miniature versions of their parents. They remained dead-still and slowly Andrew bent down and carefully lifted the two. Then, with a big smile he handed them to me. 'Congratulations, Ann,' he said proudly. 'Your first kittens – may there be many more.'

I was ecstatic – the barrier had been broken and another rare and precious species had been born in captivity. The telephone was busy that day as I informed those closely connected with the project that the rabbits had indeed produced young. Since then 10 more youngsters have been born and we now have hopes that with a few additional wild-caught rabbits we can breed up a healthy popu-

(54) *The first king cheetah born in captivity.*

(55) *Although outwardly different from the rest of the litter, 'The King'
received no special treatment from its mother, Jean.*

(56) *A young tame king cheetah
being groomed.*

(57) *King cheetah with litter-mates.*

(58) *Cheetahs devour a young nyala killed during the hailstorm.*

(59) The delicate suni antelope, bred successfully at De Wildt.

(60) A small herd of impalas flourishes at De Wildt.

(61) De Wildt's first riverine rabbit kittens.

(62) Southern Africa's most endangered
mammal, the riverine rabbit.

(63 & 64) *Howard and his team attach tracking collars to the cheetahs.*

(65) *The cheetahs, released on the farm Kempiana, with a kill.*

(66) Fire devastation in a female camp.

(67 & 68) After the fire Richard and Alan attend to burns and other injuries.

(69) The breakthrough: a full litter of king cheetah cubs born of king cheetah parents.

lation and help save an animal that man has brought to the brink of extinction. Fortunately, too, there are areas such as the Karoo National Park and farms in the Karoo region where the riverine rabbit can safely be reintroduced and will be, in the legal sense, protected.

* D. Lawson, 1986. *Ecology and Conservation of Suni in Natal.* Pietermaritzburg: Institute of Natural Resources, University of Natal.
† P. A. Basson and J. M. Hofmeyr, 1973. 'Mortalities associated with wildlife capture operations', in E. Young, ed., *The Capture and Care of Wild Animals.* Cape Town: Human & Rousseau, p.151.

CHAPTER 16

The Future

*I*n the mid 1980s the Pretoria and Cincinnati zoos agreed to the exchange of a
male king cheetah for a male white tiger. This meant that for the first time there
was to be a captive-born-and-bred king cheetah on view outside of De Wildt – a
milestone event, I thought. After all the documentation necessary for international
travel had been completed, we crated the young cat in preparation for its journey
by air to the USA. As I watched the vehicle with its precious cargo disappearing
down the farm's gravel road on its way to Jan Smuts Airport, as always I felt the
loss of one of my family. I wondered, too, whether the new owners would realize
how many years of hard work and perseverance a dedicated team had spent on
breeding the cheetah and finally, with nature's aid, had produced this unique
animal with its strikingly beautiful coat. Would this king sire cubs in the USA,
establishing a new breeding nucleus there? I knew that the North American
Regional Cheetah Studbook had been given international status and that zoological gardens and private facilities throughout the world had been requested to
register their wild-caught and captive-born populations of cheetahs. Documented
by the NOAHS Centre at the National Zoological Park in Washington, D.C., all
cheetah births, deaths, imports and transfers are noted in this annual publication
and each animal is given a stud number and record card. Fortunately, over the
years we have been able to establish a number of breeding lines at De Wildt and
from the beginning of the project our cheetahs have been named and numbered

and strict breeding records have been maintained. In future, through records published in the *International Cheetah (Acinonyx jubatus) Studbook*, we will be able to trace the breeding successes and failures of our animals that have been transferred to other institutions, and hopefully one day Lady's bloodline will be found in many parts of the world.

I thought of Lady. She was by now much too old to produce young and had joined the group of pensioners and would spend the last few years of her life at De Wildt in quiet retirement. But for the past two days Lady had eaten very little and had needed my close attention. That afternoon, after calling on the help of Shironga, I drove with him in the bakkie to her enclosure. In the past she had always come to the fence on hearing the noise of the small truck's engine but on that particular day her familiar face was not to be seen. I entered the camp with a heavy heart as I knew well that her life was drawing to a close and I feared the worst. After searching the area for a while, Shironga and I eventually found her in a patch of dense bush. As so often happens in old cheetahs and other cats, she had become thin and dehydrated, for her kidneys were failing. I sensed that she was now too weak to fight to live and that there was really nothing we could do to help her. Shironga and I returned to the hospital to fetch warm milk, minced meat and chicken but I drove back to Lady on my own as I wished to spend the time alone with her. I did not ask Dave to call – she was in no pain and I wanted her to end her life as gracefully as she had lived it.

When I approached, she lifted her head stiffly and fixed her gaze on me. Those soft eyes that had always had a sparkle in the past were now dull and sunken. Although I tried to entice her to eat the pieces of chicken and meat that I had brought, it was to no avail, and as I sat close by and watched her, I thought back over the years and pondered on her life. What a wonderful member of the family she had been! At the start of the project, while we were trying to unravel the secrets of breeding cheetahs in captivity, Lady without any trouble at all had produced a healthy litter of five. After this breakthrough, while other mothers had abandoned their cubs, Lady had possessively nurtured her young and, as if to outdo the others, the following year she had produced a litter of eight. Of these, seven had survived and all had been successfully raised by her. When her sister, Kromstert, had abandoned her cubs, there was but one cheetah mother whom we trusted to adopt and raise them – and that was Lady. And finally, in the year that we had decided not to breed at all, she had pulled out her greatest trump card and through her three daughters, Jean, Jumper and Cynthia, had unwittingly solved the age-old mystery of the rare king cheetah.

Suddenly she gave a sharp cry and slowly lifting her head she opened her deep-set eyes to look at me fixedly. Her body shuddered, her head pulled back, and then her muscles relaxed and her eyes closed for the last time. Again I felt that intense emotional bond of trust and respect that can result from a close relationship between a human and a wild animal. I had known for some time that Lady was near the end of her life, but the actual parting was awful. I could not hold back the tears that trickled freely down my face. Our wonderful years together had passed so quickly and now I had to lose a very dear friend. Her wild and independent nature, coupled with a gentleness all of her own, had made her an exceptionally easy animal to love. So aptly named, she had lived and died a Lady.

Time passed unnoticed as I sat there quietly next to her, deep in thought. The sun had dipped in the west and the peace I had experienced in the early days was with me again. As I looked at the lifeless Lady, I knew I had lost the last link with the early days of De Wildt. I knew that I had now come to the crossroads. Great changes were taking place and I had to look to the future. I was proud that in 1986, mainly as a result of our breeding successes at De Wildt, the cheetah had been taken off the list of endangered mammals in South Africa. To add to this, Dr Anton Rupert, founder of the South African Nature Foundation, had presented me with the foundation's prestigious gold medal award, thereby acknowledging my contributions to conservation. For me it was a great honour to be recognized in this way and I feel deep pride in having been given so highly coveted an award. But, as I sat there alone with Lady with the darkness creeping up on me, one troubled thought kept recurring: the fate of the free-roaming cheetah – was the species really off the endangered list? Had I really achieved anything useful at De Wildt?

The wild cheetah, I feel, is still in a very precarious position. For example, according to the latest game-count in the Kruger National Park – a protected area of 1 948 582 hectares – it is estimated that there are only 250 cheetahs. This is attributed to the animal's territorial needs for vast areas of habitat and its constant conflict with the park's large population of lions and spotted hyaenas. Attempts have been made in the past to introduce foreign cheetahs into the reserve, but the newcomers have always either been killed by the resident animals or have wandered out of the park onto surrounding farms to be shot by livestock farmers.

Perhaps if lion and spotted hyaena populations were reduced in the park the cheetah numbers would have a chance to increase. But one has to admit that the lion is a great tourist attraction – many visitors feel that their stay in the reserve is a failure if they do not see at least one of these big cats – and it is easier to spot than the elusive cheetah.

Ironically, cheetahs that live outside protected areas, on privately owned land, are classed as vermin and may legally be shot if they are found to be a danger to a farmer's livestock. To the game-farmer the cheetah is considered a worthless animal as it cannot be hunted as a trophy and it also causes damage to game stocks as it preys on the ewes and young of potentially valuable antelope. In 1973 the Washington Agreement, later known as 'CITES' or The Convention on International Trade in Endangered Species, was signed by most countries in an attempt to protect wild indigenous flora and fauna throughout the world. One of the clauses of this agreement specifies an embargo on the export of live, wild-caught spotted cats or their skins. This, however, does not apply to captive-born felids. By forbidding the movement of wild cats from one country to another, species are doomed to a stagnant status in the light of world game markets, and so illegal dealings continue to prosper. It has been suggested that the embargoes applying to the category of spotted cats be suspended and that these cats be placed instead on a controlled-hunting list. The reason behind this proposal is that if surplus animals can be legally hunted for their skins, the game-farmer will become aware of the cheetah's value as a trophy and this in turn will lead to its controlled protection. It is a sad but true state of affairs that in today's commercially orientated and overpopulated world, a wild animal will remain endangered until it becomes a marketable commodity. Thus, for better or for worse, the wild cheetah is fast becoming a rare species.

To all at De Wildt it has become apparent that, in the 1990s, because of a lack of suitable new habitats, the reintroduction of cheetahs into the wild is an almost impossible task. However, what we do know is that we have overcome most of the breeding problems encountered in captive conditions and today our ideas and findings are being copied and applied in many parts of southern Africa as well as in the rest of the world. As a result, I am confident that the cheetah can continue to be bred in captivity in years to come.

Epilogue

One hot early summer morning in 1990, a group of excited schoolchildren, guided by Miriam, wound their way on a walking trail through the centre. A few adults accompanied them and a cigarette butt or match was carelessly tossed into the long dry grass. Suddenly the air was filled with the noise of crackling flames, tearful children, panic-stricken animals and people shouting instructions. The veld had not been burnt for the past 20 years and the dead wood that remained after the hailstorm added fuel to the flames.

For five hours two fire engines, friends, zoo staff, neighbours and my own team fought a losing battle as six-metre-high flames reached for the sky and devoured everything in their path. A huge and dense mass of smoke hung over the scene. All the gates of the enclosures were quickly opened to free the frightened animals so that they could run ahead of the blaze. Some of the cheetahs could not immediately be accounted for and we could see that others were already badly burnt. The cubs sought refuge in their grass-covered shelters but these were alight and the youngsters were severely scorched. Others climbed fences in panic and disappeared into adjoining areas.

The next morning we took stock. A silent black patch of land was all that was left of the cheetah camps. Fibreglass watertanks had cracked and melted in the intense heat; fencing poles had been irreparably damaged and only a few sheets of corrugated iron remained of what had once been a series of cub shelters.

Cheetahs and brown hyaenas were found wandering in the impala camp. Bewildered, some still in a state of shock, they seemed unable to comprehend what had happened. The wild dogs were safe as they had disappeared down into their dens during the holocaust, but the vultures were not so fortunate. They had been encircled by the fire and four of the birds had died from shock and inhalation of smoke. Their recently completed man-made cliff and nesting structure was reduced to a pile of ashes.

We lost one breeding cheetah and four vultures on that sad day. Fourteen animals, including cubs, were treated for burns mainly on their torsos and paws. Their wounds required regular attention and the changing of dressings and application of medication put them through a long and painful ordeal. And to crown it all, the episode provided neighbouring staff and 'helpers' with a first-hand ground plan of the centre and as a result the perimeter fence was cut some days later and two vultures were stolen for so-called 'medicinal' purposes. It seemed so tragic that the birds should have survived the fire only to end up as fragments in a witchdoctor's shop.

I was stunned. My compassion for my family of animals was all-consuming, but I felt bitter too. That anyone could have so recklessly caused such devastation was hard to accept. In the past we had so carefully and meticulously burnt firebreaks round the perimeter fence, never dreaming that a fire could start from within. Again I was faced with the utter dejection that I had experienced before. But, as in the past, friends, supporters and family rallied round. The plight of De Wildt was publicized in the Press and also on television's '50/50' programme and once more untold encouragement helped me to realize just how important it was to carry on. Donations large and small were made, enabling us to make a start on the repairing of fences and replacing of watertanks – and obtaining a small fire unit of our own. Some of the most touching gifts came from pensioners and young schoolchildren whose letters moved me deeply.

The dramatic effect of the fire gave me cause to reassess the need for cheetah breeding. Was it really so necessary? Was the world in need of so many animal species when it could barely feed the human population? But the more I questioned, the stronger became my conviction. Yes, it all had to continue. If the cheetah had survived on this planet for over a million years, this present generation had all the more reason to preserve the cat's existence. But it would have to be a breed of cheetah that was geared to the 21st century: a breed whose survival would depend on its acceptance of a controlled environment and restricted habitat – sadly, a breed that would never be able to experience the complete freedom its ancestors had known in the past.

Select List of Research Projects Emanating from Studies Undertaken at and Associated with the De Wildt Cheetah Research Centre

1 UNPUBLISHED

Anderson, G.A. 1980. 'Communication as a social bonding mechanism in the African wild dog *Lycaon pictus*.' B.Sc.(Hons) thesis, University of Pretoria.

Bosman, A.L. 1982. 'The social behaviour of captive brown hyaenas.' B.Sc.(Hons) thesis, University of Pretoria.

Brand, D.J., Meltzer, D.G.A., Coubrough, R.I. and Van Dyk, Ann. 1975. 'Cheetah breeding.' Report of the National Zoological Gardens, Pretoria.

Degenaar, J.P. 1977. 'Aspects of reproduction in captive cheetah *Acinonyx jubatus* (Schreber, 1776).' M.Sc. thesis, University of Pretoria.

De Klerk, G.J. 1988. 'The ethology and sociology of the suni, *Neotragus moschatus*.' B.Sc.(Hons) thesis, University of Pretoria.

Dippenaar, S.M. and Ferguson, J.W.H. 1990. 'Towards a captive breeding programme for the riverine rabbit, *Bunolagus monticularis*.' Report, Department of Zoology, University of Pretoria.

Duthie, A.G. 1989. 'The ecology of the riverine rabbit *Bunolagus monticularis*.' M.Sc. thesis, University of Pretoria.

Joubert, M.J.J. 1989. 'Die seisoenale assimilasie doeltreffendheid van die aardwolf *Proteles cristatus* in gevangenskap.' B.Sc.(Hons) thesis, University of Pretoria.

Kogler, H. 1985. 'Investigations on the digestion by cheetah, (*Acinonyx jubatus*).' Report, Eugene Marais Chair in Wildlife Management, University of Pretoria.

Komen, J. 1986. 'Energy requirements and food resource of the Cape vulture *Gyps coprotheres*, in the Magaliesberg, Transvaal.' M.Sc. thesis, University of the Witwatersrand, Johannesburg.

Levitan, C.D. 1988. 'The influence of *Trinervitermes trinervoides* terpene secretions on the feeding behaviour of the aardwolf (*Proteles cristatus*), bat-eared fox (*Otocyon megalotis*), and civet (*Viverra civetta*).' B.Sc.(Hons) thesis, University of Pretoria.

Meltzer, D.G.A. 1988. 'Reproduction in the male cheetah *Acinonyx jubatus jubatus* (Schreber, 1776).' M.Sc. thesis, University of Pretoria.

Palmer, R.W. 1985. 'Investigation of the diet of *Felis caracal* and *Felis lybica* in the Karoo National Park.' B.Sc.(Hons) thesis, University of Pretoria.

Ramdohr, W.H.W. 1990. 'The effect of season, food intake and body weight on body temperature and the efficiency of digestion in captive aardwolves, *Proteles cristatus.*' B.Sc.(Hons) thesis, University of Pretoria.

Somers, M.J. 1988. 'The marking behaviour of the suni, *Neotragus moschatus.*' B.Sc.(Hons) thesis, University of Pretoria.

Van Heerden, J. 1987. 'Immobilization of wild dogs (*Lycaon pictus*) with a tiletamine hydrochloride/Zolazepam hydrochloride combination and subsequent evaluation of selected blood chemistry parameters.' Report, Faculty of Veterinary Science, Medical University of South Africa, Medunsa.

Van Heerden, J. 1981. 'The use of Medetomidine and a Medetomidine/ Ketamine hydrochloride combination in wild dogs (*Lycaon pictus*).' Report, Faculty of Veterinary Science, Medical University of South Africa, Medunsa.

Van Heerden, J. 1981. 'Electrocardiographic investigations in wild dogs (*Lycaon pictus*).' Report, Faculty of Veterinary Science, Medical University of South Africa, Medunsa.

Wilkinson, I.S. 1984. 'Chromosomes of the brown hyaena, *Hyaena brunnea.*' B.Sc.(Hons) thesis, University of Pretoria.

2 PUBLISHED

Bertschinger, H.J., Meltzer, D.G.A., Van Dyk, Ann, Coubrough, R.I., Soley, J.T., and Collett, Felicity A. 1984. 'Cheetah life-line.' *Nuclear Active.* 30: pp. 2-7.

Brand, D.J. 1980. 'Captive propagation at the National Zoological Gardens of South Africa.' *International Zoo Yearbook.* 20: pp. 107-112.

Button, C., Meltzer, D.G.A. and Mulders, M.S.G. 1981. 'The electro-cardiogram of the cheetah *Acinonyx jubatus.*' *Journal of the South African Veterinary Association.* 52(3): pp. 233-235.

Button, C., Meltzer, D.G.A., and Mulders, M.S.G. 1981.'Saffan induced poikilothermia in cheetah *Acinonyx jubatus.*' *Journal of the South African Veterinary Association.* 52(3): pp. 237-238.

Coubrough, R.I., Bertschinger, H.J. and Soley, J.T. 1978. 'Scanning electron microscopic studies on cheetah spermatozoa.' *Proceedings of the Electron Microscopy Society of Southern Africa.* 8: pp. 57-58.

Coubrough, R.I., Bertschinger, H.J., Soley, J.T. and Meltzer, D.G.A. 1976. 'Some aspects of normal and abnormal spermatozoa in cheetah (*Acinonyx jubatus*).' *Proceedings of the Electron Microscopy Society of Southern Africa.* 6: pp. 5-6.

Howard, J.G., Bush, M., Wildt, D., Brand, D.J., Ebedes, H., Van Dyk, Ann and Meltzer, D. 1981. 'Preliminary reproductive physiology studies on cheetahs in South Africa.' *Annual Proceedings of the American Association of Zoological Veterinarians.* pp. 72-74.

Meltzer, D.G.A. 1979. 'Cheetah research.' *Zoon.* 1: pp. 1-4.

O'Brien, S.J., Brand, D.J., Meltzer, D.G.A., Simmons, I.G. and Bush, M. 1981. 'The extent of genetic variation of the African cheetah (*Acinonyx jubatus*).' *Annual Proceedings of the American Association of Zoological Veterinarians.* pp. 74-77.

O'Brien, S.J., Roelke, M.E., Marker, L., Newman, A., Winkler, C.A., Meltzer, D., Colly, L., Evermann, J.F., Bush, M. and Wildt, D.E. 1985. 'Genetic basis for species vulnerability in the cheetah.' *Science.* 227: pp. 1428-1434.

O'Brien, S.J., Wildt, D.E., Goldman, D., Merril, C.R. and Bush, M. 1983. 'The cheetah is depauperate in genetic variation.' *Science.* 221: pp. 459-462.

Pettifer, H.L. 1981. 'The experimental release of captive-bred cheetah (*Acinonyx jubatus*) into the natural environment,' in J.A. Chapman and D. Pursley, eds, *The Worldwide Furbearer Conference Proceedings. Falls Church, Virginia: R.R. Donnelley, pp. 1001-1024.*

Soley, J.T. and Coubrough, R.I. 1981. 'A "pouch-like" defect in the basal plate region of cheetah spermatozoa.' *Proceedings of the Electron Microscopy Society of Southern Africa.* 11: pp. 121-122.

Van Aarde, R.J. and Skinner, J.D. 1986. 'The pattern of space use by relocated servals, *Felis serval.' African Journal of Ecology.* 24: pp. 97-101.

Van Aarde, R.J. and Skinner, J.D. 1987. 'Range use by brown hyaenas *Hyaena brunnea* relocated into an agricultural area of the Transvaal, South Africa.' *Journal of Zoology, London.* 212: pp. 350-352.

Van Aarde, R.J. and Van Dyk, Ann. 1986. 'Inheritance of the king coat colour pattern in cheetahs *Acinonyx jubatus.' Journal of Zoology, London.* 209: pp. 573-578.

Van Heerden, J. 1981. 'The role of integumental glands in the social and mating behaviour of the hunting dog *Lycaon pictus* (Temminck, 1820).' *Onderstepoort Journal of Veterinary Research.* 48: pp. 19-21.

Van Heerden, J. 1986. 'Disease and mortality of captive wild dogs *Lycaon pictus.' South African Journal of Wildlife Research.* 16: pp. 7-11.

Van Heerden, J., Dauth, J., Komen, J. and Myer, E. 1987. 'The use of ketamine hydrochloride in the immobilization of the Cape vulture, *Gyps coprotheres.' Journal of the South African Veterinary Association.* 58(3): pp. 143-144.

Van Heerden, J. and De Vos, V. 1981. 'Immobilization of the hunting dog *Lycaon pictus* with ketamine hydrochloride and a fentanyl/droperidol combination.' *South African Journal of Wildlife Research.* 11: pp. 112-113.

Van Heerden, J., Komen, J. and Myer, E. 1987. 'Serum biochemical and haematological parameters in the Cape vulture *Gyps coprotheres.' Journal of the South African Veterinary Association.* 58(3): pp. 145-146.

Van Heerden, J. and Kuhn, F. 1985. 'Reproduction in captive hunting dogs *Lycaon pictus.' South African Journal of Wildlife Research.* 15: pp. 80-84.

Van Heerden, J., Swart, W.H. and Meltzer, D.G.A. 1980. 'Serum antibody levels before and after administration of live canine distemper vaccine to the wild dog *Lycaon pictus*.' *Journal of the South African Veterinary Association*. 51(4): pp. 283-284.

Wildt, D.E., Bush, M., Howard, J.G., O'Brien, S.J., Meltzer, D., Van Dyk, Ann, Ebedes, H. and Brand, D.J. 1983. 'Unique seminal quality in the South African cheetah and a comparative evaluation in the domestic cat.' *Biology of Reproduction*. 29: pp. 1019-1025.

Wildt, D.E., Chakraborty, P.K., Meltzer, D. and Bush, M. 1984. 'The pituitary and gonadal response to LH releasing hormone administration in the female and male cheetah.' *Journal of Endocrinology*. 101: pp. 51-56.

Wildt, D.E., Meltzer, D., Chakraborty, P.K. and Bush, M. 1984. 'Adrenal-testicular and pituitary relationships in the cheetah subjected to anesthesia/electroejaculation.' *Biology of Reproduction*. 30: pp. 665-672.

Wilkinson, I.S. and Skinner, J.D. 1988. 'Efficacy of 22Na turnover in ecophysiological studies of carnivores.' *South African Journal of Zoology*. 23: pp. 32-36.

Wrogemann, N. 1975. *Cheetah Under the Sun*. Johannesburg: McGraw-Hill.

INDEX

Plate numbers are indicated by figures in **bold** type.